Springer Theses

Recognizing Outstanding Ph.D. Research

Aims and Scope

The series "Springer Theses" brings together a selection of the very best Ph.D. theses from around the world and across the physical sciences. Nominated and endorsed by two recognized specialists, each published volume has been selected for its scientific excellence and the high impact of its contents for the pertinent field of research. For greater accessibility to non-specialists, the published versions include an extended introduction, as well as a foreword by the student's supervisor explaining the special relevance of the work for the field. As a whole, the series will provide a valuable resource both for newcomers to the research fields described, and for other scientists seeking detailed background information on special questions. Finally, it provides an accredited documentation of the valuable contributions made by today's younger generation of scientists.

Theses are accepted into the series by invited nomination only and must fulfill all of the following criteria

- They must be written in good English.
- The topic should fall within the confines of Chemistry, Physics, Earth Sciences, Engineering and related interdisciplinary fields such as Materials, Nanoscience, Chemical Engineering, Complex Systems and Biophysics.
- The work reported in the thesis must represent a significant scientific advance.
- If the thesis includes previously published material, permission to reproduce this must be gained from the respective copyright holder.
- They must have been examined and passed during the 12 months prior to nomination.
- Each thesis should include a foreword by the supervisor outlining the significance of its content.
- The theses should have a clearly defined structure including an introduction accessible to scientists not expert in that particular field.

Indexed by zbMATH.

More information about this series at http://www.springer.com/series/8790

Álvaro Díaz Fernández

Reshaping of Dirac Cones in Topological Insulators and Graphene

Doctoral Thesis accepted by
Universidad Complutense de Madrid,
Madrid, Spain

Springer

Author
Dr. Álvaro Díaz Fernández
Escuela Técnica Superior de
Arquitectura de Madrid
Universidad Politécnica de Madrid
Madrid, Spain

Supervisors
Prof. Francisco Domínguez-Adame Acosta
Facultad de Ciencias Físicas
Universidad Complutense de Madrid
Madrid, Spain

Prof. Elena Díaz García
Facultad de Ciencias Físicas
Universidad Complutense de Madrid
Madrid, Spain

ISSN 2190-5053 ISSN 2190-5061 (electronic)
Springer Theses
ISBN 978-3-030-61557-4 ISBN 978-3-030-61555-0 (eBook)
https://doi.org/10.1007/978-3-030-61555-0

This Springer imprint is published by the registered company Springer Nature Switzerland AG
The registered company address is: Gewerbestrasse 11, 6330 Cham, Switzerland

A great deal of my work
is just playing with equations
and seeing what they give.

—PAUL A. M. DIRAC

A Marta, siempre

Supervisors' Foreword

The following thesis by Dr. Álvaro Díaz Fernández was performed at the Materials Physics Department of the Complutense University of Madrid from 2016 to 2019. The subject of research addressed by Álvaro falls into the new field of topological condensed matter physics. This thesis is definitely an outstanding piece of work in terms of physical significance and the clarity of writing.

Symmetry-protected topological phases of matter are gapped phases which cannot be adiabatically connected to the vacuum without breaking the symmetries. A particularly relevant example is that of topological insulators. These cannot be connected to ordinary band insulators without closing the gap unless time-reversal symmetry is broken. The most striking manifestation of the time-reversal symmetry protection of the topologically insulating phase is that, when a finite material is considered, gapless surface states with well-defined helicities emerge. Moreover, such gapless excitations are Dirac cones, thereby setting topological insulators as an example of Dirac quantum matter. Some of their characteristics are predicted to have a strong impact on transport properties at the nanoscale. Therefore, a deep understanding of such characteristics is crucial for future functioning devices.

With this in mind, Álvaro pursued a clear goal during his Ph.D. research: analyse the effect of external fields on the robustness of the exotic properties of topological insulators and search for different approaches to manipulate them in favour of quantum transport. With such a well established but not straightforward objective, he was able to prove that an electric field perpendicularly applied to the surface of a topological insulator preserves the Dirac cones, while altering the Fermi velocity as a function of the field's intensity. Moreover, Álvaro extended the scope of his research to graphene, a Dirac material of great relevance, to show that this phenomenon could also be observed in metallic armchair nanoribbons. This seems to be a clear indication that the effect should be observable in other Dirac materials as well. As a particular realization of the proposal, Álvaro proposed the usage of the built-in field created by a δ-layer of impurities near the surface. This proved to be particularly relevant to show that the surface states where robust and would coexist with a Rashba-split two-dimensional electron gas.

In order to explore the effects of a magnetic field, Álvaro considered different orientations. Although a magnetic field breaks time-reversal symmetry, if applied with a particular orientation it may preserve the mirror symmetry of the Hamiltonian. This is particularly relevant when considering dual topological insulators, that is, those which display both topologically insulating and topological crystalline insulating behaviours. The latter are topological insulators protected by crystal symmetries. If time-reversal symmetry is broken but mirror symmetry is not, the topological state will not deform into a trivial state. In combination with electric fields, Álvaro could show that the dispersion relation is greatly altered without gapping out the surface states.

Finally, Álvaro concluded his Ph.D. research by extending his analysis into the realm of periodically driven topological insulators. These can be addressed by means of Floquet theory. As he could show by considering the full three-dimensional Hamiltonian, there is an interplay between bulk and surface states which leads to similar behaviours as with the static case. However, there is an extra handle in the polarization of the applied field. This can lead to time-reversal symmetry breaking, as occurs for circularly polarized fields, or no symmetry breaking, as with linearly polarizing fields. In the former case, a gap opens up in the quasienergy spectrum, whereas in the latter there is no gap opening. Additionally, by tuning the field strength, the slope of the Dirac cones could also be altered, in similarity to the static situation.

Álvaro made an invaluable effort to build a bridge connecting the field of topology and symmetries in theoretical physics to the relevance of topological materials in condensed matter physics. His thesis begins with a clear accessible account for postgraduate students and non-expert researchers of symmetry protected topological phases and its relevance for quantum transport at the nanoscale. Indeed, it reveals as a crucial reference textbook to fill up the gap between the academic graduate knowledge and the basics needed for an introduction into research studies on this field. We are truly delighted to have Álvaro as one of our Ph.D. students and to witness that he has become a first-class researcher.

Madrid, Spain
June 2020

Francisco Domínguez-Adame Acosta
Elena Díaz García

Abstract

Quantum mechanics was in all its splendour at the beginning of the twentieth century with the great minds of Schrödinger, Heisenberg, Dirac, Pauli and many others. The concepts introduced by then were truly revolutionary, even more than Einstein's relativity, one would dare say. One may have a feeling that nowadays we only exploit the consequences of quantum mechanics to build devices such as transistors or light-emitting diodes, which on the other hand have changed the world we live in beyond imagination. With regard to the theoretical front, it may seem that one only has to turn the mathematical handle to work out consequences of quantum mechanics, although no new concepts are in sight. However, nothing could be farther from the truth. We are privileged to live in what has been dubbed as a *second quantum revolution*. It is the era of entanglement. One may argue that entanglement comes from the old era with Einstein's attempts to show the apparent inconsistencies of quantum mechanics. However, it is now that we are starting to understand entanglement and are exploring its far-reaching consequences. A particularly relevant example is that of topological quantum matter. If entanglement is supplemented with symmetries, a new class of phases emerges known as *symmetry protected topological phases*. In such phases, topological behaviour occurs whenever certain symmetries are preserved. Topological insulators are a paradigmatic example of symmetry protected topological phases. The gapped ground state of these materials is degenerate with that of ordinary band insulators. However, both systems belong to different topological sectors, which is observed macroscopically on the fact that the former displays edge or surface states. These edge or surface excitations are actually very special: they behave like massless Dirac fermions with well-defined helicities. The peculiar dispersion of topological insulators is shared by other so-called Dirac materials, the most prominent of which is graphene.

On the other hand, it is well known that the Fermi velocity plays a crucial role in quantum transport and by manipulating it one can reshape the transport properties of a bare system. In ordinary semiconductors, one only needs to shift the Fermi energy to achieve such a manipulation. However, this is not the case in a Dirac cone spectrum since the velocity is the same everywhere (it is the slope of the cone). In order to exploit the full power of these novel materials, it becomes interesting to be able to have control on the Fermi velocity. Additionally, it is interesting to study how robust topological surface states are against disorder, which naturally occurs in real systems upon cleavage and later exposure to the environment. The purpose of this thesis is two-fold: dynamically tune the Fermi velocity by using external fields and observe the robustness against a thin layer of impurities at the interface between a topological and a trivial insulator. Previous to embarking on such a journey, an extensive chapter is included to cover basic ideas of the systems under study. The thesis is organized in seven chapters, the first one being a brief historical introduction and the last one a short set of conclusions. The content of the remaining five chapters could be briefly summarized in the following ideas:

- **Chapter 2: Two-Band Models**

  A fairly extensive account of basic ideas of topology and models built by means of symmetry arguments. In this chapter, we set the foundations for the following chapters, which represent the core of the thesis.

- **Chapter 3: Reshaping of Dirac Cones by Electric Fields**

  By applying uniform electric fields to topological surface states and metallic graphene armchair nanoribbons, we are able to change the Fermi velocity in a fully dynamical and experimentally feasible way. The topological protection of the Dirac cones is discussed, along with hand-waving arguments that relate to the quantum-confined Stark effect, allowing us understand the physics behind these results.

- **Chapter 4: Reshaping of Dirac Cones by Magnetic Fields**

  Magnetic fields are known to break time-reversal symmetry and represent the destruction of the topological insulating phase. However, for specific orientations of the field that preserve crystalline symmetries such as mirror symmetry, the topological signature of surface states can survive the magnetic field. As we are able to observe, a suitably oriented magnetic field, complemented with an electric field perpendicular to the surface, renders the Dirac cones anisotropic, thereby leading to an anisotropic Fermi velocity.

- **Chapter 5: Surface States in δ-doped Topological Boundaries**

  Experiments have demonstrated that the topological surface states are protected regardless of impurities, as long as these are non-magnetic. Moreover, they have been shown to coexist with a Rashba-split two-dimensional electron gas, that occurs due to structural inversion asymmetry and a built-in electric field due to the impurities. In this chapter, we propose a method to achieve further control of this effect by evaporating a δ-layer of donor atoms during growth. In this case, the impurity concentration can be controlled in a very precise fashion. As we will show by means of an exactly solvable model, the Dirac state indeed survives the δ-layer, while coexisting with a Rashba-split two-dimensional electron gas with non-trivial spin textures.

- **Chapter 6: Floquet Engineering of Dirac Cones**

  The use of periodic drivings to achieve greater control of quantum states is becoming more and more relevant. Indeed, it provides with a level of tunability that is far beyond the limits of the static case. In this thesis, we propose to apply external ac fields to a topological boundary and graphene. Although such scenarios have already been discussed in the literature by using surface effective Hamiltonians and perturbation theory, we show that even with the bulk states the Dirac cones in the quasienergy spectrum remain robust. Moreover, features from the static regime are also observed in this case, with the extra handle of polarization. In particular, a change in the slope of the cones is observed.

In summary, this thesis aims at exploring basic, fundamental properties of topological surface states and Dirac cones when exposed to different perturbations. We expect to add value to the unprecedented quantum transport properties of topological insulators for their future implementation in devices. More importantly, we expect to unravel and confirm properties that are ascribed to these materials by exploring the most elementary situations, such as applying uniform electric and magnetic fields.

Publications Related to This Thesis

Articles

1. **Quantum-confined Stark effect in band-inverted junctions**
   A. Díaz-Fernández and F. Domínguez-Adame
   Physica E **93**, 230 (2017)

2. **Tuning the Fermi velocity in Dirac materials with an electric field**
   A. Díaz-Fernández, L. Chico, J. W. González and F. Domínguez-Adame
   Scientific Reports **7**, 8058 (2017)

3. **Electric control of the bandgap in quantum wells with band-inverted junctions**
   A. Díaz-Fernández, L. Chico and F. Domínguez-Adame
   Journal of Physics: Condensed Matter **29**, 475301 (2017)

4. **Robust midgap states in band-inverted junctions under electric and magnetic fields**
   A. Díaz-Fernández, N. del Valle and F. Domínguez-Adame
   Beilstein Journal of Nanotechnology **9**, 1405 (2018)

5. **Topologically protected states in δ-doped junctions with band inversion**
   A. Díaz-Fernández, N. del Valle, E. Díaz and F. Domínguez-Adame
   Physical Review B **98**, 085424 (2018)

6. **Floquet engineering of Dirac cones on the surface of a topological insulator**
   A. Díaz-Fernández, E. Díaz, A. Gómez-León, G. Platero and
   F. Domínguez-Adame
   Physical Review B **100**, 075412 (2019)

The following publication has evolved from this doctoral dissertation:

1. **Inducing anisotropies in Dirac fermions by periodic driving**
   A. Díaz-Fernández
   Journal of Physics: Condensed Matter **32**, 495501 (2020)

Other Publications

1. **Electric field manipulation of surface states in topological semimetals**
   Y. Baba, A. Díaz-Fernández, E. Díaz, F. Domínguez-Adame and
   R. A. Molina
   Physical Review B **100**, 165105 (2019)

Conference Contributions

1. **Tuning the Fermi velocity in band-inverted junctions**
   A. Díaz-Fernández, L. Chico, J. W. González and F. Domínguez-Adame
   Talk at *XIV GISC Workshop*
   Madrid (Spain), 27 January 2017

2. **Tailoring Fermi's velocity in topological insulators by an electric field**
   A. Díaz-Fernández, L. Chico, J. W. González and F. Domínguez-Adame
   Talk at *Trends in Nanotechnology*
   Dresden (Germany), 5–9 June 2017

3. **Interface states in band-inverted junctions under electric and magnetic fields**
   N. del Valle, A. Díaz-Fernández and F. Domínguez-Adame
   Poster at *WE-Heraeus-Physics School on Exciting nanostructures*
   Hamburg (Germany), 16–21 June 2017

4. **Two-dimensional Dirac materials with tunable Fermi velocity**
   A. Díaz-Fernández, L. Chico, J. W. González and F. Domínguez-Adame
   Poster at *Topological Matter School*
   San Sebastián (Spain), DIPC, 21–25 August 2017

5. **Two-dimensional Dirac materials with tunable Fermi velocity**
   A. Díaz-Fernández and F. Domínguez-Adame
   Poster at *14th European Conference on Molecular Electronics*
   Dresden (Germany), 29 August–2 September 2017

6. **Field control of surface states in topological materials**
   A. Díaz-Fernández, N. del Valle and F. Domínguez-Adame
   Talk at *20th International Conference on Superlattices, Nanostructures and Nanodevices*
   Madrid (Spain), 23–27 July 2018

7. **Topologically protected surface states in the presence of a two-dimensional electron gas**
   A. Díaz-Fernández, N. del Valle, E. Díaz and F. Domínguez-Adame
   Poster at *Summer School on Collective Behaviour in Quantum Matter*
   Trieste (Italy), ICTP, 27 August–14 September 2018

8. **Topologically protected surface states in the presence of a two-dimensional electron gas**
   A. Díaz-Fernández, N. del Valle, E. Díaz and F. Domínguez-Adame
   Poster at *Topological phases in condensed matter and cold atom systems*
   Corsica (France), IES Cargèse, 1–13 October 2018

9. **Controlling topologically protected states by external fields and doping**
A. Díaz-Fernández, E. Díaz and F. Domínguez-Adame
Talk at *XVI GISC Workshop*
Madrid (Spain), 11 January 2019

10. **Controlling topologically protected states by external fields and doping**
A. Díaz-Fernández, E. Díaz and F. Domínguez-Adame
Poster at *Anyons in Quantum Many-Body Systems*
Dresden (Germany), 21–25 January 2019

11. **Controlling topologically protected states by external fields and doping**
A. Díaz-Fernández, E. Díaz and F. Domínguez-Adame
Talk at *APS March Meeting*
Boston (United States), 4–8 March 2019

12. **Floquet engineering of Dirac cones on the surface of a topological insulator**
A. Díaz-Fernández, E. Díaz, A. Gómez León, G. Platero and F. Domínguez-Adame
Talk at *Dynamic Dirac Quantum Matter*
Florida (United States), 16–20 December 2019

Seminars and Lectures given by Invitation

1. **Quantum Spin Hall Effect**
A. Díaz-Fernández
Lectures at *Departamento de Física Fundamental. Facultad de Ciencias, Universidad de Salamanca*
Salamanca (Spain), 27–29 June 2017

2. **Negative Differential Resistance in InAs/GaSb Core-Shell Nanowires and the Effect of High Magnetic Fields**
A. Díaz-Fernández
Seminar at *Dipartimento di Fisica. Università degli Studi di Pavia*
Pavia (Italy), 10 October 2017

3. **Can we tune surface states in topological materials?**
A. Díaz-Fernández
Seminar at *Institut für Physik. Johannes Gutenberg-Universität Mainz*
Mainz (Germany), 17 July 2018

Acknowledgements

This is by far the most complicated part of a thesis. Indeed, it is the part that everyone reads even if the thesis is uninteresting to them. Moreover, I have struggled with choosing the language: should I write it in Spanish, English or !Xóõ? The last one seemed far too easy, the first one was a tough one, so I decided to go with the one in between: English. I shall be inconsistent anyway and switch to Spanish at some point. It is also too complicated to decide who to acknowledge. I will therefore leave some people out, please do not take it personally. Quoting the King of Rock 'n' Roll: you were always on my mind.

At the risk of being predictable, I would like to start by thanking my supervisors, Prof. Francisco Domínguez-Adame and Dr. Elena Díaz. Each of you deserves far more than a few lines, so I will try my best. I promise my words are genuine. Francisco, it has been 6 years since we first met. It may not seem like a lot of time, but it is. I vividly remember the day you told me about the possibility of working in your group. At the time I was a student in your Solid State Physics class and, to be fair, I was not interested at all in anything related to the subject. As most of my fellow students interested in Theoretical Physics, I was aiming to pursue a career in Particle Physics or something related to black holes and things like that. I am most grateful to you for showing me the beauty of Condensed Matter Physics. Today, we work together to share our passion for the field by organizing outreach sessions on Condensed Matter Physics, by discussing better ways to make students aware of this field, by learning together. I am extremely grateful to the freedom you have given me, as well as to always being so supportive of my initiatives. As I said at the beginning, it may seem like 6 years is not a lot of time, but I bet that very few students have spent more time with their supervisor than I have. We even shared an office during our stay at the University of Warwick! If you do not mind my saying this, I like to think that you are now much more than a supervisor. Rather, I think of you as a friend. I will not extend any longer for not making this too lengthy, but there is no end to my sincere thankfulness. Elena, I met you also some time ago, although I had not have the chance to do so properly until 3 years ago. Since then, I have experienced endless support throughout the thesis, not only on the scientific side but also on the personal one. You have been caring and understanding at all times. More importantly, you have always treated me as equal, always being keen to learn about a topic that was not so much related to your previous research. I truly appreciate it. Muchísimas gracias por todo.

Among my collaborators, I wish to express my most sincere gratitude to Dr. Leonor Chico. One of the core papers of this thesis is related to a collaboration we started some time ago. I am indebted to you for the idea of extending to the realm of graphene our studies related to applying electric fields upon topological surface states. It proved to be genuinely interesting. I want to thank the people in Salamanca who have been so kind since we first met about 4 years ago. Especially, I want to extend my gratitude to Dr. Enrique Diez and Vito Clericò for hosting me at the Universidad de Salamanca to give a series of lectures on the quantum spin Hall effect, as well as for their support on the research performed during my stay in Pavia. Next, I would like to acknowledge the work of Natalia del Valle, now a Master student, who collaborated with us in two of the papers that are included in this thesis. Thank you for your time and I really wish you all the best of luck. Lastly, I would like to thank Dr. Gloria Platero and Dr. Álvaro Gómez-León, whom I am having the opportunity of collaborating with on the last part of this thesis devoted to Floquet theory. Many thanks for sharing your insight and intuition into this problem and for being considerate with my being a newbie in this area.

I would like to acknowledge those academics at foreign institutions whom I have had the pleasure to meet and collaborate with. I will begin with Prof. Rudolf Römer, who gave me the chance of being part of his research group at the University of Warwick about four years ago. That was my first experience at a foreign institution and I cherish very good memories, all thanks to your being so welcoming. You even introduced me to your family at a farewell barbecue at your home when my stay was coming to an end. Thank you very much and I really look forward to seeing you soon. I would also like to thank Dr. Vittorio Bellani. Always in a good mood, I finally managed to learn some Italian thanks to you and to grasp the importance of Experimental Physics. Those 3 months at the Università di Pavia were incredibly memorable and I could even have the chance of performing some experiments (we did some Raman spectroscopy). Thank you so much for making my stay a really smooth one and for taking me to discover hidden regions of Liguria. I would also like to thank your sincere interest in Marta, always being kind and caring towards her. Grazie mille. Finally, I would like to thank Dr. Matteo Rizzi for his generous invitation to the JGU Mainz to work on the topology of optical lattices and artificial gauge fields. I enjoyed our conversations about physics and hope that we can collaborate again sometime in the near future.

These years I have also been lucky to share my lunchtimes with a nice crowd: Álvaro, Antonio, Bea, Belén Sotillo, Fer, Francisco, Gregor, Jaime, Javi, Jesús, María, Marina, Mica, Paloma, Ruth and Víctor. Please forgive me if I forget a single name (which is very likely), but there are a lot of you. I would like to thank a few people in particular. Manuel. Thank you for being so generous, where would I be without the turrones de Jijona every winter and the nísperos de Alicante every spring. I hugely enjoy our conversations about education and life, I believe we have a lot in common and hope that our joint effort will lead to the next revolution in Physics education. Although I still think that teaching multiplication tables is useless and that children should be taught Hilbert spaces instead. However, we share more things apart from our ideas about Physics and Education, we also go to the theatre together and we are the biggest fans of Jaime Urrutia. Yuriko. I met you when teaching Materials Physics. You were the only one who would come to the blackboard to attempt (with success) a solution to the problems that had been proposed. This was about 3 years ago I think. Right now, we have deep conversations where we learn quite a bit of Physics and I hope it will continue this way. We also discuss sometimes life and music, although concerning music I listen more than I say. Thank you very much for sharing your passion for piano music and for inviting me to your concerts. Above all, I would like to thank you for always having a smile on your face, despite the huge loads of work you have to do. Félix. You are always keen to learn about my progress now that you are not at the office anymore, and I really appreciate that. I want to thank you for those delicious chocolates that you brought me every time you came from a trip to Finland. You also did not mind to devote some of your time to give me some tips about Greece that were key to organizing my honeymoon. Thank you. Marta. We met a while ago, although it is recently that we are becoming closer. I always feel so cheered up by our conversations about the successful future that awaits us after we finish the thesis. I am sure that we will find some nice place where they will host us, preferably a decent one. Fer. Thank you for being so kind and welcoming when I arrived at the department, we have had some good times at the office and you are deeply missed. Although your Old Testament is still there to remind me of you. Ali and Belén Cortés. You were also really welcoming since the beginning, as well as caring and kind-hearted. Ali, thank you so much for being my chief provider of delicious eco vegetables. I will always envy your prosperous vegetable garden. I wish all of you the best of luck and would really like to keep in touch.

A special mention goes to the great moments with Anna (el Rey), Anto, Edo (la Reina) and Edu: the funniest and most unproductive times of my Ph.D., and probably those where I have eaten more chocolate than ever. Siempre nos quedarán los Chupa Chups de Pamplona! Seriously, it has been 4 years now since we met and I always think of when will be our next gathering. I am really looking forward to seeing you, it has been more than a year now. I hope it will be soon, I would love to join again to cook some Italian dish (Edu and I would chop onions and peppers only, but I still enjoy it very much).

My buddy and big brother Andreas, I cannot forget you here and you deserve a special place. Since San Sebastián, we have seen each other quite a few times, two in the warm winter of Dresden. I think that I have been extremely lucky to have met you, I learn a lot when we discuss Physics and life, especially doing so surrounded by the vineyards of Oppenheim. I had a lot of fun in Brussels eating mussels in a funny-tasting sauce at the Lobster House followed by a delicious waffle at the Australian place and a beer at Delirium Café. I hope this friendship will last forever.

Philipp, you are unbeatable when doing a handstand at the beach in Cargèse, what a nice spot to attend some lectures!

The people at Cambridge: Max, Ollie, Stephen and Attila. We had such a great time in Trieste. I will always remember the Scala Dirac, the morning walk through the park and those cheeky swims at the jetty while enjoying the beautiful sunset. I would like to see you soon to cook some noodles, have a capo in bi and wrap up the day by playing Cabo!

Finalmente, mis queridos Paulo y Miguel. Nos conocimos haciendo la carrera y seguimos en contacto. Nuestra amistad es muy fuerte, sois verdaderamente importantes para mí. Lamento que no nos veamos con mayor frecuencia, pero sé que podremos contar los unos con los otros. Siempre nos quedarán las comidas de Casa Dani y los paseos por el Retiro.

We come to an end here. Well, not really. This project could not have been possible without the support of my family. A mis padres. Mamá, gracias por tu confianza y apoyo incondicional, por transmitirme calma y sosiego, por tu capacidad de superación y reinvención. Papá, gracias por respetar siempre mis decisiones, aun cuando te parecían disparatadas. A ambos, gracias por la libertad que me habéis concedido y las oportunidades que me habéis brindado siempre.

A mis abuelos y mi tía Aida. Gracias por vuestro apoyo constante y por confiar en mí against all odds. Abuela, gracias por nuestras conversaciones sobre literatura, cocina, teatro, cine… No paro de aprender de tí, gracias por enseñarme tanto, hablar contigo es un placer. Adoro tu cocina, ya sabes que tus tartas de moca y los merenguitos son mi perdición. Tía Aida, gracias por ser mi faro en la distancia, por todos tus consejos y por tu humor ante todo en la vida. Eres un referente para mí. Abuelo, sabes que te quiero como a nadie, gracias por el amor que siempre me has dado. Contigo tengo los mejores recuerdos de mi infancia.

Agradezco a mi abuelo Rafael el transmitirnos la pasión por el trabajo y su afán por el conocimiento. A mis tíos, Ana, Blanca, David, Javier y Sergio, por vuestro enorme cariño. A falta de hermanos, a mis primos, Ana, Amalia, Carlota, Gonzalo, Javi, Luci, María, Marta, Nico, Sergio y Silvia, con los que me lo paso como un enano.

A Martha, por acompañarme durante mi infancia y mi adolescencia y por enseñarme a atarme los cordones.

Agradezco a Marta, Enrique, Borja y Encarna el quererme y acogerme desde el principio como uno más. Son ya casi doce años desde que nos conocemos. Marta y Enrique, sois maravillosos. Nunca olvidaré las conversaciones con Enrique sobre temas existenciales, poesía, política. Gracias por transmitirme tu sabiduría. Marta, gracias por el cariño y la confianza que siempre has tenido conmigo, por tratarme como a un hijo. Borja, me has hecho llorar de risa, gracias por ser tan íntegro y honesto siempre. Encarna, no puedo olvidar todo tu cariño desde el principio, tu humor, tu fortaleza, los veranos en la Sierra y esa comida española que es la más rica que he probado en mi vida (junto con la de mi abuela, claro).

Por encima de todo, a Marta. Cualquier cosa que diga se queda corta, no es posible resumir en un párrafo todo lo que siento. Has sido la persona que más ha confiado en mí, tu amor es infinito y puro. He aprendido tanto de tí que yo ya no soy yo, soy los dos. Siento que esta tesis forma parte de nuestro proyecto común, de doce años que llevamos trabajando juntos para ser cada día mejores. Esta tesis es tuya. Gracias por estar a mi lado. Te quiero. Siempre.

Contents

Chapter 1
Introduction

Condensed Matter Physics is by far the vastest field of Physics. The amount of phenomena it encompasses is in fact overwhelming. Its success relies on the ability to make astoundingly accurate predictions about the world around us. Not only that, it has led to major breakthroughs that have changed the world we live in, the most conspicuous example possibly being the transistor.

In the last few decades, a truly abstract field of Mathematics is permeating all of modern Condensed Matter Physics: Topology. It is now mainstream to read the words *Topological Matter, Topological Phase Transitions* and other concepts alike. The field started with the pioneering works of David J. Thouless, F. Duncan M. Haldane and J. Michael Kosterlitz, who shared the Nobel Prize in Physics 2016 for their theoretical discoveries on this topic. These seminal contributions took place around the 1980's. Although they had an impact on the community at the time, it was the discovery of topological insulators in 2004 by Charles Kane and Eugene Mele that really boosted the field. Since then, the area of topological matter has increased exponentially. The Nobel Lecture by Haldane is a delightful read about the evolution and some key discoveries in this field [1].

This introduction is not aimed to provide a detailed account of the field of topological matter. The reader is referred to Refs. [2–10] for excellent reviews on the subject. Instead, we shall present a few details of the historical development that led to the materials of our interest: topological insulators and graphene. We will be brief, however, since the first chapter of this Thesis consists on a detailed account of the models and the topological features that underly these systems. Moreover, we shall consider only single-particle physics, since it will be the regime of interest in this Thesis. Without further ado, let us start.

The story begins in February 5th, 1980 at the High Magnetic Field Laboratory in Grenoble. At two in the morning, a young Klaus von Klitzing was performing measurements in silicon MOSFET[1] devices in an attempt to understand how to improve the mobility of electrons in such devices [11]. What he found was radically

[1]Metal-oxide-semiconductor field-effect transistor.

Á. Díaz Fernández, *Reshaping of Dirac Cones in Topological Insulators and Graphene*, Springer Theses, https://doi.org/10.1007/978-3-030-61555-0_1

shocking. Upon applying strong magnetic fields to the samples and performing Hall measurements, he discovered that the Hall resistance did not follow what Drude's theory predicted. Instead of increasing linearly with magnetic field, it developed a series of plateaus where it remained constant for a range of fields. Moreover, the Hall conductance was quantized in multiples of e^2/h, with e the elementary charge and h Planck's constant [12]. This result is truly astonishing: it is the perfect quantization of a macroscopic quantity. Additionally, when the Hall resistance sat on a plateau, the longitudinal resistance dropped to zero, a sign of dissipationless transport. Von Klitzing was awarded the Nobel Prize in Physics 1985 for the discovery of the quantum Hall effect.

Only a year later, Robert B. Laughlin provided a truly elegant argument to explain the physics of the quantum Hall effect that relied on gauge invariance and static disorder [13]. Later that year, Bertrand I. Halperin [14] realized that Laughlin's argument was incomplete or, rather, that it relied on some assumptions that where key to the understanding of the effect and that had not been pointed out by Laughlin. Most notably, Halperin pointed out the existence of edge states in the mobility gap that arises due to static disorder. These states formed one-dimensional chiral wires that carried current dissipationlessly and where responsible for the sharp quantization of the conductance. Without knowing it, Halperin had found one of the first signatures of topological matter: edge states.

As we said, the physics of the quantum Hall effect relied very much on the presence of static disorder and the presence of edges. However, as Thouless and coworkers pointed out in 1982, such arguments seem to leave a gap to the understanding of why the conductance is quantized irrespective of the geometry of the sample. It seemed that such a quantization should stem from the bulk material. Dealing with arbitrary disorder was, however, rather tricky. To circumvent such a problem, Thouless proposed the usage of a periodic potential in a square lattice threaded by a perpendicular magnetic field. The electron energy levels of such a system had already been obtained in 1976 by Douglas R. Hofstadter [15] while doing his Ph. D. Thesis. Such energy levels as a function of the flux threading a unit cell form a fractal structure known as the Hofstadter butterfly. Using the Kubo formula and the Hofstadter model as a prototypical example, Thouless was indeed able to show that the conductivity was quantized in integer multiples of e^2/h [16]. The next year, Barry Simon [17] made a connection with the recently discovered Berry phase by Michael V. Berry [18], the integer obtained by Thouless and coworkers and the mathematical theory of topology. He noted that the integer found by Thouless and collaborators was the first Chern number. This number is what is called a topological invariant. Topological invariants are truly useful in topology to be able to decide whether two objects can be continuously connected or not [19]. As a result, they are insensitive to details and provide information about global properties of such objects. The paradigmatic example is the difference between a sphere and torus. Locally they are very similar; globally they are completely different, as it can be told by the latter having a hole and the former not having one. In this case, the number of holes, also known as the genus, is a topological invariant. Since the Chern number is a topological invariant, it cannot change under small perturbations. As a result, the quantized conductivity must also be protected by topology. In order to be able to define the Chern number in the quantum Hall scenario, however, it was shown that the Fermi energy must lie within an energy gap or a mobility gap for that matter. That is, the system had to be an insulator in the bulk. When bulk extended states are found at the Fermi energy, the Chern number becomes ill-defined and the conductivity is not quantized anymore. Upon pushing the Fermi energy back again into an energy gap, the conductivity becomes quantized once more [20]. The transition from one plateau to the next where the Chern number changes is known as a topological phase transition.

A few years later, in 1988, Haldane made a seminal contribution [21] where he showed that the only true requirement for obtaining a nontrivial quantum Hall conductance was to break time-reversal symmetry. It is only then that the Chern number is nonzero. In his model, which consisted on a sheet of graphene (at the time known as a two-dimensional single sheet of graphite), Haldane proposed a tight-binding model with complex next-nearest-neighbor hoppings. As had been shown by Philip R. Wallace [22], graphene showed a semimetallic spectrum where the conduction and valence bands touch at isolated points of the Brillouin zone. By breaking time-reversal symmetry, Haldane was able to open up a gap and render the system an insulator, one of the key requirements to observing quantum Hall physics. In this scenario, Haldane obtained that the conductivity was indeed quantized. Moreover, if such a model is solved on a strip instead of being solved in the bulk, chiral edge states appear whenever time-reversal symmetry is broken. This is one of the first clear connections between topology and edge state excitations. In 1993, Yasuhiro Hatsugai established such a connection formally, leading to what is known as the bulk-boundary correspondence [23].

Although there were a lot more developments in the years to follow in the field of topological matter, we shall jump directly to the ones that are most interesting to this Thesis. We come straight to 2004. That year, Charles L. Kane and Eugene J. Mele from Penn University made a contribution that would boost the field of topological matter [24, 25]. They realized that spin-orbit interaction in graphene, although small, would lead to two time-reversed copies of the Haldane model, one for each spin. In this case where time-reversal symmetry is unbroken, the Chern number is identically zero and one does not find a quantized Hall conductivity. However, each spin subspace shows an opposite non-zero Chern number and their difference is therefore nonzero. This implies that one can obtain a quantized spin Hall conductivity response. This argument can be generalized to non-conserving spin terms and the resulting topological invariant is commonly referred to as a $\mathbb{Z}_2$ invariant since it is a binary quantity. Taking into account that different spin subspaces are time-reversed Haldane partners, at the edges one finds helical spin currents. The system found by Kane and Mele is commonly referred to as a quantum spin Hall insulator and it is the first example of a topological insulator.

An experimental observation of the quantum spin Hall effect in bare graphene is very complicated. This is due to the fact that spin-orbit interactions are very small in graphene due to the lightness of carbon [3]. B. Andrei Bernevig, Taylor L. Hughes and Shou-Cheng Zhang [26] proposed that the same effect would occur by employing heavier elements where spin-orbit interactions where stronger. In particular, they considered HgTe. However, HgTe is a semimetal in bulk. One way to open an energy gap is to grow nanostructures to reduce the symmetry of the system, so they proposed using a quantum well. Their seminal contribution showed that such a system indeed hosted the same physics as that proposed by Kane and Mele a year earlier. Crucially, only one year later the group of Laurens W. Molenkamp in Würzburg was able to perform measurements on HgTe quantum wells [27], confirming the predictions made by Bernevig, Hughes and Zhang.

By then, it was clear that the quantum spin Hall effect was a time-reversal symmetric quantum Hall analog of phenomena that occurs in two-dimensional systems. However, there is more to the story. Based on the works done in 1997 by Alexander Altland and Martin R. Zirnbauer on the classification of random matrices [28], Andreas P. Schnyder, Shinsei Ryu and collaborators [8, 29–31], and Alexei Kitaev [32] proposed what is commonly referred to as the ten-fold way or periodic table of topological insulators and superconductors. In this periodic table, ten symmetry classes based on the presence or absence of fundamental symmetries (time-reversal, particle-hole and chiral symmetries) are considered. As these researchers were able to show, depending on the symmetry class and the spatial dimension, the ground states of gapped phases of single-particle Hamiltonians obeying the symmetries of such a class could be characterized by a topological invariant. Such topological invariants could be trivial, meaning that all ground states within that class were topologically equivalent, or they could be $\mathbb{Z}$ or $\mathbb{Z}_2$ numbers, meaning that there where a countably infinite number of classes or only two distinct classes, respectively. The gapped phases of the symmetry class where the quantum Hall effect belongs to had $\mathbb{Z}$ invariants only in even dimensions. In all other cases, such phases would be trivial and no quantum Hall phase is expected to be observed in odd dimensions. Hence, there can be no three-dimensional quantum Hall effect.[2] On the other hand, the symmetry class where the quantum spin Hall effect belongs to showed non-trivial $\mathbb{Z}_2$ invariants both in two and three dimensions. Hence, it was expected to find three-dimensional analogs of the quantum spin Hall effect with topological surface states. In 2008, the first three dimensional topological insulator was observed by means of angle-resolved photoemission spectroscopy in $Bi_{1-x}Sb_x$ compounds [33]. The review by M. Zahid Hasan and C. L. Kane provides great detail on the history of this material and the experiments associated to the detection of its topological surface states [3]. However, we are more interested in what are called *second generation materials* [34–36] since they can be described with the models utilized in this Thesis, whereas the former cannot. These include Bi_2Se_3, Bi_2Te_3 and Sb_2Te_3, three materials that are well-known for their thermoelectric properties. The surface states of these materials have massless Dirac-like dispersions. Moreover, their energy gap is rather larger, making them suitable for exploiting their topological properties even at room temperature. By using symmetry arguments, one can see that these materials can be modeled by means of a $3 + 1$ Dirac equation, as is shown in the first chapter of this Thesis. Finally, it has

[2]There can be three-dimensional quantum Hall states in theory, although these would be obtained by stacking two-dimensional layers of the quantum Hall state and there is no topological protection in such a case.

been demonstrated that the Altland-Zirnbauer classification could be enriched by adding crystalline symmetries. As a result, materials that would be trivial according to such a classification can be nontrivial by displaying symmetries such as mirror symmetry. Although less robust than regular three-dimensional topological insulators, these crystalline topological insulators [7, 37, 38] have also been extensively investigated and experiments have unraveled their existence. More importantly to us, the model behind these systems is also a $3 + 1$ Dirac equation.

On a different note, even though graphene has already been mentioned when discussing the Haldane and Kane–Mele models, it is also interesting to make a few notes on aspects related to this Thesis. An excellent review of graphene can be found in Ref. [39] for further details. It was already known that graphene, even without spin-orbit or time-reversal-symmetry-breaking perturbations, could host edge states depending on the edge termination [40, 41]. Armchair nanoribbons displayed no edge states and the spectrum was very size-dependent, which could render the nanoribbons metallic or semiconducting. In the former case, the low energy excitations showed massless Dirac-like spectra. Zigzag and bearded nanoribbons, in contrast, hosted non-dispersive edge states, irrespective of the system size, suggesting that these should stem from some kind of bulk-edge correspondence. The works of Hatsugai and Ryu [42] showed that this was indeed the case and that the edge states of zigzag and bearded nanoribbons where topological in origin. As a result, metallic armchair nanoribbons can be gapped out quite easily by adding perturbations. However, as was pointed out by Juergen Wurm and collaborators [43], a low energy description exhibits topological properties related to a hidden pseudovalley structure, and therefore no gap openings are expected to occur for sufficiently smooth space-dependent potentials.

Before we conclude, we would like to point out a few motivating aspects of this Thesis, although we shall not be exhaustive and leave most details to the introduction of each chapter. As the title of this Thesis suggests, we are mostly interested in manipulating the Dirac cones of topological insulators and graphene. One of the key properties that is related to quantum transport is the Fermi velocity, which in these systems corresponds merely to the slope of the cone. Hence, in contrast to regular semiconductors, by simply shifting the Fermi level one does not change the Fermi velocity. In this Thesis, we propose to dynamically modify such a velocity by applying external fields. In graphene, there exist several proposals towards modifying such defining property [44–49]. However, none of these proposals allow to alter the Fermi velocity in a dynamical way. For instance, the Fermi velocity can be altered by interaction with a substrate. However, this substrate is fixed and, in order to change the Fermi velocity, one would need to change the substrate. In our proposal, the Fermi velocity becomes field dependent and, therefore, it can tuned on the fly. The potential interest of modifying the Fermi velocity is the great impact it has on quantum transport measurements. As an example, it has been proposed that a sheet of graphene on top of a patterned substrate, which would in turn create a patterned configuration of velocities, can lead to a complete suppression of the transmission [50].

In any case, even if applications are absent or far in the future, we may conclude this introduction by quoting Duncan Haldane on a telephone interview that followed the announcement of the 2016 Nobel Prize in Physics:

It's very difficult to know whether something is useful or not, but one can know that it's exciting.

References

1. Haldane FDM (2017) Nobel lecture: topological quantum matter. Rev Mod Phys 89:040502
2. König M, Buhmann H, Molenkamp LW, Hughes T, Liu C-X, Qi X-L, Zhang S-C (2008) The quantum spin Hall effect: theory and experiment. J Phys Soc Jpn 77:031007
3. Hasan MZ, Kane CL (2010) Colloquium: topological insulators. Rev Mod Phys 82:3045
4. Qi X-L, Zhang S-C (2011) Topological insulators and superconductors. Rev Mod Phys 83:1057
5. Yan B, Zhang S-C (2012) Topological materials. Rep Prog Phys 75:096501
6. Ando Y (2013) Topological insulator materials. J Phys Soc Jpn 82:102001
7. Ando Y, Fu L (2015) Topological crystalline insulators and topological superconductors: from concepts to materials. Annu Rev Condens Matter Phys 6:361
8. Chiu C-K, Teo JCY, Schnyder AP, Ryu S (2016) Classification of topological quantum matter with symmetries. Rev Mod Phys 88:035005
9. Yan B, Felser C (2017) Topological materials: Weyl semimetals. Annu Rev Condens Matter Phys 8:337
10. Wen X-G (2017) Colloquium: zoo of quantum-topological phases of matter. Rev Mod Phys 89:041004
11. Doucot B, Pasquier V, Duplantier B, Rivasseau V (2005) The quantum Hall effect: Poincaré seminar 2004. Birkhäuser, Basel
12. Klitzing KV, Dorda G, Pepper M (1980) New method for high accuracy determination of the fine-structure constant based on quantized Hall resistance. Phys Rev Lett 45:494
13. Laughlin RB (1981) Quantized Hall conductivity in two dimensions. Phys Rev B 23:5632
14. Halperin BI (1982) Quantized Hall conductance, current-carrying edge states, and the existence of extended states in a two-dimensional disordered potential. Phys Rev B 25:2185
15. Hofstadter DR (1976) Energy levels and wave functions of Bloch electrons in rational and irrational magnetic fields. Phys Rev B 14:2239
16. Thouless DJ, Kohmoto M, Nightingale MP, den Nijs M (1982) Quantized Hall conductance in a two-dimensional periodic potential. Phys Rev Lett 49:405
17. Simon B (1983) Holonomy, the quantum adiabatic theorem, and Berry's phase. Phys Rev Lett 51:2167
18. Berry MV (1984) Quantal phase factors accompanying adiabatic changes. Proc R Soc Lond A 392:45
19. Nakahara M (2003) Geometry, topology and physics. Taylor & Francis, Boca Raton
20. Thouless DJ (1998) Topological quantum numbers in nonrelativistic physics. World Scientific, Singapore
21. Haldane FDM (1988) Model for a quantum Hall effect without Landau levels: condensed-matter realization of the parity anomaly. Phys Rev Lett 61:2015
22. Wallace PR (1947) The band theory of graphite. Phys Rev 71:622
23. Hatsugai Y (1993) Chern number and edge states in the integer quantum Hall effect. Phys Rev Lett 71:3697
24. Kane CL, Mele EJ (2005) Z_2 topological order and the quantum spin Hall effect. Phys Rev Lett 95:146802
25. Kane CL, Mele EJ (2005) Quantum spin Hall effect in graphene. Phys Rev Lett 95:226801

26. Bernevig BA, Hughes TL, Zhang S-C (2006) Quantum spin Hall effect and topological phase transition in HgTe quantum wells. Science 314:1757
27. Köonig M, Wiedmann S, Brüne C, Roth A, Buhmann H, Molenkamp LW, Qi X-L, Zhang S-C (2007) Quantum spin Hall insulator state in HgTe quantum wells. Science 318:766
28. Altland A, Zirnbauer MR (1997) Nonstandard symmetry classes in mesoscopic normal-superconducting hybrid structures. Phys Rev B 55:1142
29. Schnyder AP, Ryu S, Furusaki A, Ludwig AWW (2008) Classification of topological insulators and superconductors in three spatial dimensions. Phys Rev B 78:195125
30. Schnyder AP, Ryu S, Furusaki A, Ludwig AWW (2009) Classification of Topological Insulators and Superconductors. In: AIP conference proceedings, vol 1134, p 10
31. Ryu S, Schnyder AP, Furusaki A, Ludwig AWW (2010) Topological insulators and superconductors: tenfold way and dimensional hierarchy. New J Phys 12:065010
32. Kitaev A (2009) Periodic table for topological insulators and superconductors. In: AIP conference proceedings, vol 1134, p 22
33. Hsieh D, Qian D, Wray L, Xia Y, Hor YS, Cava RJ, Hasan MZ (2008) A topological Dirac insulator in a quantum spin Hall phase. Nature 452:970
34. Moore J (2009) The next generation. Nat Phys 5:378
35. Xia Y, Qian D, Hsieh D, Wray L, Pal A, Lin H, Bansil A, Grauer D, Hor YS, Cava RJ, Hasan MZ (2009) Observation of a largegap topological-insulator class with a single Dirac cone on the surface. Nat Phys 5:398
36. Zhang H, Liu C-X, Qi X-L, Dai X, Fang Z, Zhang S-C (2009) Topological insulators in Bi_2Se_3, Bi_2Te_3 and Sb_2Te_3 with a single Dirac cone on the surface. Nat Phys 5:438
37. Fu L (2011) Topological crystalline insulators. Phys Rev Lett 106:106802
38. Hsieh TH, Lin H, Liu J, Duan W, Bansil A, Fu L (2012) Topological crystalline insulators in the SnTe material class. Nat Commun 3:982
39. Castro Neto AH, Guinea F, Peres NMR, Novoselov KS, Geim AK (2009) The electronic properties of graphene. Rev Mod Phys 81:109
40. Brey L, Fertig HA (2006) Electronic states of graphene nanoribbons studied with the Dirac equation. Phys Rev B 73:235411
41. Wurm J, Wimmer M, Adagideli I, Richter K, Baranger HU (2009) Interfaces within graphene nanoribbons. New J Phys 11:095022
42. Ryu S, Hatsugai Y (2002) Topological origin of zero-energy edge states in particle-hole symmetric systems. Phys Rev Lett 89:077002
43. Wurm J, Wimmer M, Richter K (2012) Symmetries and the conductance of graphene nanoribbons with long-range disorder. Phys Rev B 85:245418
44. Li G, Luican A, Lopes dos Santos JMB, Castro Neto AH, Reina A, Kong J, Andrei EY (2009) Observation of Van Hove singularities in twisted graphene layers. Nat Phys 6:109
45. Trambly de Laissardi'ere G, Mayou D, Magaud L (2010) Localization of Dirac electrons in rotated graphene bilayers. Nano Lett 10:804
46. Hicks J, Sprinkle M, Shepperd K, Wang F, Tejeda A, Taleb- Ibrahimi A, Bertran F, Le F'evre P, de Heer WA, Berger C, Conrad EH (2011) Symmetry breaking in commensurate graphene rotational stacking: comparison of theory and experiment. Phys Rev B 83:205403
47. Hwang C, Siegel DA, Mo S-K, Regan W, Ismach A, Zhang Y, Zettl A, Lanzara A (2012) Fermi velocity engineering in graphene by substrate modification. Sci Rep 2:590
48. Elias DC, Gorbachev RV, Mayorov AS, Morozov SV, Zhukov AA, Blake P, Ponomarenko LA, Grigorieva IV, Novoselov KS, Guinea F, Geim AK (2011) Dirac cones reshaped by interaction effects in suspended graphene. Nat Phys 7:701
49. Miao L, Wang ZF, Ming W, Yao M-Y, Wang M, Yang F, Song YR, Zhu F, Fedorov AV, Sun Z, Gao CL, Liu C, Xue Q-K, Liu C- X, Liu F, Qian D, Jia J-F (2013) Quasiparticle dynamics in reshaped helical Dirac cone of topological insulators. Proc Natl Acad Sci 110:2758
50. Lima JRF, Pereira LFC, Bezerra CG (2016) Controlling resonant tunneling in graphene via Fermi velocity engineering. J Appl Phys 119:244301

Chapter 2
Two-Band Models

The Dirac equation [1] is commonly referred to as the relativistic equation for the electron [2]. Although this view is compelling since the equation satisfies Lorenz covariance, it leads to the correct energy-momentum relation for the electron and it incorporates spin in a very natural way, it has its drawbacks. In particular, the equation should correspond to a single particle equation, that for the electron. However, in order to be stabilized, it requires the ad hoc introduction of a Dirac sea that is completely full. In order to circumvent this problem, quantum field theory must be used and the wavefunction in the Dirac equation is to be reinterpreted rather as a quantum field [3].

In contrast to its counterpart in High Energy Physics, in Condensed Matter Physics the Dirac equation arises as a low energy description of Dirac materials such as graphene or topological insulators. Therefore, there is no such problem as to giving an interpretation for an unbounded spectrum from below since the model is only valid for low energies. Also, the problem of an infinite negative charge due to the completely full Dirac sea is not present, not only because of the previous argument but also because the ionic charge neutralizes this effect. The final purpose of this chapter will be to derive the Dirac equation that is encountered in the two systems considered in this Thesis: topological insulators and graphene. Typically this is done by means of what is known as $\boldsymbol{k} \cdot \boldsymbol{p}$ theory [4–7]. However, we shall approach the subject from the so-called *method of invariants*, which states that the Hamiltonian has to be invariant under the symmetries of the the system [7–10]. Before we tackle that problem, it is worth to describe briefly what topology is and its relation to symmetries, as we do in the following section.

Á. Díaz Fernández, *Reshaping of Dirac Cones in Topological Insulators and Graphene*, Springer Theses, https://doi.org/10.1007/978-3-030-61555-0_2

2.1 Topology and Symmetries

In this section, we will provide a brief motivation to the concepts of topology and symmetry, that will prove useful to our understanding in the next sections. In particular, we will introduce the concept of *topological invariant* and why it is important in two specific physical applications.

2.1.1 Homotopy and Winding Numbers

Topology is the branch of mathematics that deals with continuity. An important concept in topology is that of *homeomorphism*. Loosely speaking, a homeomorphism is a transformation from an object[1] to another such that it preserves the global properties of the former and vice versa. The canonical example is that where a coffee cup is continuously deformed into a doughnut. These two are said to be homeomorphic or topologically equivalent. However, it is usually somehow difficult to find if two objects are homeomorphic. Rather, we make use of the so-called *topological invariants*. These are quantities that we can compute and compare: if the topological invariant of one object is different from that of another, then the two objects are not homeomorphic [11]. Notice that the converse need not be true, that is, if two objects share a few topological invariants then they might not be homeomorphic. This is because, up to date, there is no knowledge of all the topological invariants that exist and, therefore, it is not possible to compare them all. In the case of the coffee cup and the doughnut, the topological invariant is the number of holes. Then, we can conclude that an orange an a doughnut cannot be homeomorphic since they have a different number of holes.

In the quantum world, there are also topological invariants to compute. Most of the times, one can understand these invariants by analysing the winding in the phase of the wavefunction around special points of the parameter space. In order to understand the following statements a little better, we shall give a short introduction to *homotopy*, but we shall restrict to a basic introduction, leaving the mathematical details to specific books on the subject [11]. Let us consider some topological space, Y. Examples of topological spaces are the real line $\mathbb{R}$, or more generally, the Euclidean space $\mathbb{R}^n$, the circle S^1, the torus $\mathbb{T} = S^1 \times S^1$, and many more examples. We can view paths in Y as continuous mappings from some other topological space X into Y, that is, a path is a continuous mapping $\alpha : X \to Y$. If two such paths can be deformed into one another, they are said to be *homotopic* or *homotopically equivalent*. As a result, we can define equivalence classes known as *homotopy classes* denoted by $[\alpha]$, where all paths equivalent to the path α fit in. Given two paths α and β, such that the end point of α coincides with the starting point of β, we may construct a product path, $\gamma = \alpha\beta$. In the case where X is a segment $X = [0, 1]$ or the unit circle S^1, this is achieved by simply pasting the starting point of β to the final point of α. This product naturally

[1] More formally, a topological space.

translates into a product between equivalent classes,

$$[\alpha][\beta] = [\alpha\beta] . \tag{2.1.1}$$

Let $\{X, Y\}$ be the set of all homotopy classes of maps from X to Y. That set, together with the multiplication law above, satisfies the four axioms of a group. If $X = S^n$ is the nth sphere, then such group is called the nth homotopy group of Y, $\pi_n(Y)$. For $n = 1$, this is also known as the *fundamental group*, $\pi_1(Y)$. Particularly interesting is to find which groups are isomorphic to the homotopy groups, that is, those groups that can be mapped one-to-one to the homotopy groups. Let us consider for instance $Y = \mathbb{R}^n$. Since all loops in $\mathbb{R}^n$ are contractible to a point, all loops belong to the same equivalence class. Therefore, there is only one element in $\pi_n(\mathbb{R}^m)$, which means that $\pi_n(\mathbb{R}^m)$ is trivial, $\pi_n(\mathbb{R}^m) \cong \{e\}$, where e is the identity element. We say that $\mathbb{R}^n$ is *simply connected*. On the other hand, if $Y = S^n$, then we find that $\pi_n(S^n) \cong \mathbb{Z}$. That is, the nth homotopy group of the nth sphere is isomorphic to the group of integers under addition. We can understand this easily by considering the $n = 1$ case. Each equivalence class can be identified by an integer, namely, the number of times the loops in that class wind around the circle, say m times. We call such an integer a *winding number*. It is then clear that the product of two equivalence classes with winding numbers m and n lead to the equivalence class of winding number $m + n$. Therefore, $\mathbb{Z}$ has the same structure as $\pi_1(S^1)$ and, therefore, they are isomorphic. The same applies to $\pi_n(S^n)$. We can see that the winding numbers provide us with an interesting result, namely, that two loops of different winding numbers belong to disjoint classes and, therefore, they cannot be continuosly deformed into one another. Hence, the winding number is our first encounter with a *topological invariant*.

Let us consider two examples where the winding number arises: the Aharonov–Bohm effect [3, 12] and the Su-Schrieffer–Heeger model [13]. We shall consider $\hbar = 1$ hereafter. Let us start with the Aharonov–Bohm effect, which consists on the following: imagine a charged particle moving in free space. Now, place an infinite solenoid of radius R along the Z-axis, carrying a steady current which, in turn, generates a nonzero magnetic field inside the solenoid, $\boldsymbol{B}$. Outside, the magnetic field is zero. However, the vector potential $\boldsymbol{A}$ is not zero outside the solenoid. Indeed, in order to have

$$\Phi = \oint_C \boldsymbol{A} \cdot \mathrm{d}\boldsymbol{r} , \tag{2.1.2}$$

where Φ is the flux threading the solenoid and C is a path enclosing the latter, we must have in cylindrical coordinates

$$\boldsymbol{A} = \frac{\Phi}{2\pi r}\hat{\boldsymbol{e}}_\phi , \tag{2.1.3}$$

in the region outside the solenoid, $r > R$. Notice that $\nabla \times \boldsymbol{A} = 0$ outside the solenoid, as it should be because $\boldsymbol{B} = 0$ in that region. The Hamiltonian for a charged particle in the background of a gauge field $\boldsymbol{A}$ is given by

$$H = \frac{1}{2m}\left(\boldsymbol{p} - q\boldsymbol{A}\right)^2 , \tag{2.1.4}$$

where m is the mass of the particle and q is its charge, $\boldsymbol{p}$ is the momentum operator. We can remove the vector potential by means of a gauge transformation of the form

$$\psi(\boldsymbol{r}) \to \exp\left(\mathrm{i}\, q \int^{r} \boldsymbol{A} \cdot \mathrm{d}\boldsymbol{r}'\right) \psi(\boldsymbol{r}) , \tag{2.1.5}$$

where the integral is along any path. In the field-free region where the charged particle is allowed to move, we can write $\boldsymbol{A} = \nabla\lambda$, where

$$\lambda(\phi) = \frac{\Phi}{2\pi}\phi . \tag{2.1.6}$$

We can also write this equation in a slightly more cumbersome manner which will nevertheless prove to be useful afterwards

$$\lambda(\phi) = \frac{\Phi}{2\pi\,\mathrm{i}} \log\left[\eta(\phi)\right] , \qquad \eta(\phi) = \exp(\mathrm{i}\,\phi) . \tag{2.1.7}$$

Then, the phase factor acquired along a trajectory starting at ϕ_i and finishing at ϕ_f would be

$$\gamma = -\mathrm{i}\frac{\Phi}{\Phi_0} \log\left[\frac{\eta(\phi_f)}{\eta(\phi_i)}\right] , \tag{2.1.8}$$

where $\Phi_0 = 2\pi/q$ is the Dirac flux quantum. If the path is a closed loop, then γ corresponds to a phase difference and it affects an interference measurement. In that case, $\phi_f = \phi_i + 2\pi N$, where N is the number of times the path winds around the solenoid and, therefore,

$$\frac{\gamma_{\mathrm{AB}}}{2\pi} = N\frac{\Phi}{\Phi_0} . \tag{2.1.9}$$

In this calculation, the actual shape of the path is irrelevant, as long as it encloses the solenoid N times. This phase is known as the *Aharonov–Bohm phase*. Notice that if Φ is an integer multiple of Φ_0, the system is gauge invariant, since γ would be an integer multiple of 2π and, therefore, the interference pattern resulting from having or not a flux would remain the same. Related to our discussion about topology, N is the winding number. Indeed, it arises from considering the mappings

$$\begin{aligned} \eta \,:\, S^1 &\to U(1) \\ \phi &\mapsto \eta(\phi) = \exp(\mathrm{i}\,\phi) . \end{aligned} \tag{2.1.10}$$

Since the group space of $U(1)$ is S^1, then these mappings are characterized by the fundamental group $\pi_1(S^1) \cong \mathbb{Z}$. In this case, the winding number can be given the geometrical interpretation that we have found of the number of times a path winds around the solenoid. This is because the configuration space is $\mathbb{R}^2 - \{0\}$, that is, a plane with a hole at the position of the solenoid, whose first homotopy group is also $\mathbb{Z}$. We say that both topological spaces, $\mathbb{R}^2 - \{0\}$ and S^1, are of the same *homotopy type*.

The second example will be the Su-Schrieffer–Heeger (SSH) model. This is a model for polyacetylene, a polymer of carbon and hydrogen atoms, $(CH)_x$. Atomic theory predicts that the ground state for carbon is $2s^2 2p^2$. However, when forming compounds, carbon excites to $2s^1 2p^3$ so that the four unpaired orbitals can hybridize. For instance, one can form four sp^3 orbitals where the carbon atom sits in the middle of a tetrahedron. That way, it is possible to maximize the overlap between orbitals when forming a crystal lattice which, as a result, compensates the increase in energy for carbon being excited in the first place. In this case, the resulting system is diamond. These strong covalent σ-bonds are responsible for most of the mechanical properties of diamond. Also, since there are no free electrons to roam around the crystal, the system is an insulator. However, there are also other possible hybridizations, most notably the sp^2 hybridization. Here, the carbon atom sits at the center of a triangle, each of the three sp^2 orbitals pointing from that position to each vertex of the triangle. The extra unhybridized p orbital is perpendicular to such triangle. With this arrangement, we can form an array of carbon atoms by overlapping one sp^2 orbital from one carbon atom to another sp^2 orbital from another carbon atom. The third sp^2 orbital is then bonded to hydrogen. The perpendicular p orbitals weakly couple to each other forming π bonds. Since all p orbitals form π bonds with each other, the result is a π band. Each carbon atom donates a single electron to that band and, therefore, it is half-filled because the band is degenerate in spin. The system is then expected to be metallic. However, as Rudolf Peierls pointed out, this situation is not possible at low temperatures [14, 15]. Instead, he proposed the lattice to be distorted, reducing the discrete translational symmetry of the lattice. Let every other atom come closer to its neighbour, so that the strength of the overlap between neighbouring atoms becomes staggered. As a result of this symmetry reduction, the unit cell now contains two atoms instead a single atom and, therefore, the Brillouin zone is halved. The effect of the perturbation in the nearly free electron model is to open up a gap at the boundaries of the Brillouin zone. The net result is that all electrons fully occupy a π band that is lower in energy with respect to the original π band, consequently reducing the energy of the system. Thus, the system becomes an insulator. There are details to this model that are not to our interest but can be consulted in the original reference by SSH [13]. For our purposes, suffices to say that we can consider a tight-binding model with staggered hoppings for the p orbitals [16].

Let us denote the two atoms in the unit cell by A and B. In the following, we shall measure distances in units of the lattice constant, a. The distance between atoms in the unit cell will be $0 < \delta < 1$. Therefore, positions of the A atoms will be given by $R_A = m$ and those of the B atoms will be given by $R_B = m + \delta$. See Fig. 2.1.

On each atom, we have a localized p orbital, $\phi(x - R_\alpha) = \langle x | R_\alpha \rangle$. For convenience of calculations, we shall consider that these orbitals form an orthonormal basis, that is, $\langle R_\alpha | R'_\beta \rangle = \delta_{\alpha\beta}\delta_{RR'}$. Also, for convenience of notation, we can write $|R_\alpha\rangle \equiv |m, \alpha\rangle = |m\rangle \otimes |\alpha\rangle$. We can then build any state in that basis as follows

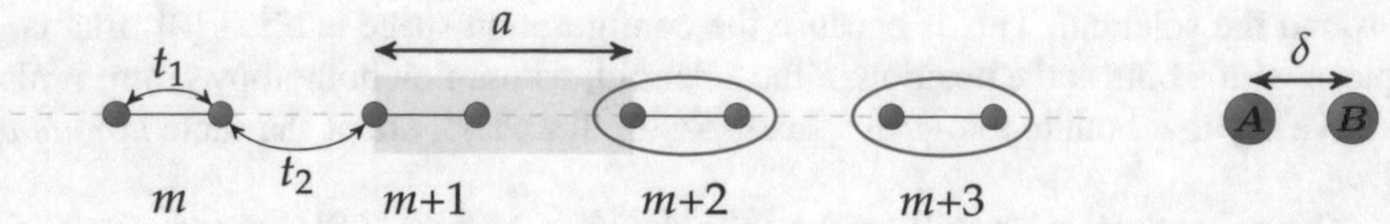

Fig. 2.1 SSH chain. The lattice constant is $a = 1$ and the separation between A and B atoms within a blue oval is δ. The hoppings are staggered and take values t_1 and t_2. A unit cell is shown in yellow

$$|\psi\rangle = \left(\frac{1}{\sqrt{N}}\sum_n c_n|n\rangle\right) \otimes \left(\frac{1}{\sqrt{2}}\sum_\alpha \varphi_\alpha|\alpha\rangle\right), \tag{2.1.11}$$

where N is the total number of unit cells and $c_n, \varphi_\alpha \in \mathbb{C}$. We assume hereafter that $|\psi\rangle$ is normalized. The Hamiltonian we shall consider is a nearest-neighbour tight-binding Hamiltonian

$$H = \sum_n \left[t_1|n, A\rangle\langle n, B| + t_2|n+1, A\rangle\langle n, B| + \text{h.c.}\right], \tag{2.1.12}$$

where the intracell hopping is t_1 and the intercell hopping is t_2 and we take them to be real, which can always be done for one-dimensional systems, and h.c. denotes the Hermitian conjugate. Here we have fixed the onsite energy to zero for convenience. We can equivalently write this Hamiltonian as follows

$$H = t_1 \mathbb{1}_N \otimes \sigma_x + t_2\left[T_1 \otimes \sigma_+ + T_{-1} \otimes \sigma_-\right], \tag{2.1.13}$$

where $\mathbb{1}_d$ is the d-dimensional identity matrix and we have introduced the translation operators in the lattice subspace

$$T_d = \sum_n |n+d\rangle\langle n|, \tag{2.1.14}$$

and the lowering and raising operators in the cell subspace

$$\sigma_\pm = \frac{1}{2}\left(\sigma_x \pm \mathrm{i}\,\sigma_y\right), \tag{2.1.15}$$

where σ_i with $i = x, y, z$ are the three standard Pauli matrices. This Hamiltonian is manifestly translationally invariant and, as such, we can choose $|\psi\rangle$ to be an eigenstate of the translation operators. This in turn allows us to introduce a quantum number, the quasimomentum k, such that the coefficients in the lattice expansion, c_n, are given by $c_n = \exp(\mathrm{i}\,kn)$. That is,

$$|\psi_k\rangle = \left(\frac{1}{\sqrt{N}}\sum_n e^{\mathrm{i}kn}|n\rangle\right)\otimes\left(\frac{1}{\sqrt{2}}\sum_\alpha \varphi_\alpha|\alpha\rangle\right) \equiv |\phi_k\rangle\otimes|\mu\rangle\,. \tag{2.1.16}$$

Clearly, $|\psi_k\rangle$ is an eigenstate of $T_d \otimes \mathbb{1}_2$ with eigenvalue $\exp(-\mathrm{i}kd)$. Therefore, if we apply H to $|\psi_k\rangle$ it is straightforward to obtain

$$H|\psi_k\rangle = |\phi_k\rangle\otimes\left[t_1\sigma_x + t_2\left(e^{-\mathrm{i}k}\sigma_+ + e^{\mathrm{i}k}\sigma_-\right)\right]|\mu\rangle\,. \tag{2.1.17}$$

The term in brackets is easily computed

$$\left[t_1\sigma_x + \frac{t_2}{2}\left(e^{-\mathrm{i}k}\sigma_+ + e^{\mathrm{i}k}\sigma_-\right)\right]|\mu\rangle = \frac{1}{\sqrt{2}}\left[\delta^*(k)\varphi_B|A\rangle + \delta(k)\varphi_A|B\rangle\right], \tag{2.1.18}$$

where we have defined

$$\delta(k) = t_1 + t_2\exp(\mathrm{i}\,k) = |\delta(k)|\exp[\mathrm{i}\,\theta(k)]\,. \tag{2.1.19}$$

Therefore, we finally find that

$$\langle\psi_k|H|\psi_k\rangle = \frac{1}{2}\left[\delta^*(k)\varphi_A^*\varphi_B + \delta(k)\varphi_B^*\varphi_A\right]\,. \tag{2.1.20}$$

We will now make use of the variational theorem, which states that the optimum φ_α are those that minimize the energy. That is, we must minimize

$$E(k) = \frac{\langle\psi_k|H|\psi_k\rangle}{\langle\psi_k|\psi_k\rangle}, \tag{2.1.21}$$

with respect to φ_α^*. Since

$$\langle\psi_k|\psi_k\rangle = \frac{1}{2}\left[\varphi_A^*\varphi_A + \varphi_B^*\varphi_B\right], \tag{2.1.22}$$

applying the above prescription we find that

$$\begin{pmatrix} 0 & \delta^*(k) \\ \delta(k) & 0 \end{pmatrix}\begin{pmatrix}\varphi_A \\ \varphi_B\end{pmatrix} = E(k)\begin{pmatrix}\varphi_A \\ \varphi_B\end{pmatrix}\,. \tag{2.1.23}$$

As it is now, it is fairly easy to obtain the dispersion relation and the amplitudes by diagonalizing the matrix above, which we shall denote by the Bloch Hamiltonian $H(k)$ hereafter,

$$E_s(k) = s|\delta(k)|\,, \qquad \boldsymbol{\chi}_s(k) \equiv \begin{pmatrix}\varphi_A(k) \\ \varphi_B(k)\end{pmatrix}_s = \begin{pmatrix} s e^{-\mathrm{i}\theta(k)} \\ 1\end{pmatrix}, \tag{2.1.24}$$

where $s = \pm 1$. Notice that, unless $t_1 = t_2$, the spectrum possesses an energy gap given by $2|t_1 - t_2|$. The eigenstates, however, are defined up to an arbitrary phase $\exp[\mathrm{i}\,\phi(k)]$. If chosen to be such that $\phi(k) = \theta(k)/2$, the eigenstates above are written as follows

$$\boldsymbol{\chi}_s(k) = \mathcal{R}\,[\theta(k)]\,\boldsymbol{\chi}_s^0\,, \tag{2.1.25}$$

where

$$\mathcal{R}\,[\theta(k)] = \exp\left(-\mathrm{i}\,\frac{\theta(k)}{2}\sigma_z\right)\,, \qquad \boldsymbol{\chi}_s^0 = \begin{pmatrix} s \\ 1 \end{pmatrix}\,. \tag{2.1.26}$$

The matrix $\mathcal{R}\,[\theta(k)]$ resembles that of a spin rotation about the Z-axis of angle $\theta(k)$ [17]. Just like with spin, a rotation of 2π leads to a minus sign. Equivalently, the state acquires a phase of π upon $\theta(k)$ completing a full rotation. This is very similar to the Aharonov–Bohm effect, where the mapping $\eta(\phi) = \exp(\mathrm{i}\,\phi)$ is now given by a more complicated mapping $\exp[\mathrm{i}\,\theta(k)]$ that is not simply given by $\exp(\mathrm{i}\,k)$. However, it is still a mapping from S^1 (the Brillouin zone is periodic) to $U(1)$. Therefore, we can introduce once again the notion of winding numbers, which we shall denote by ν. Nevertheless, it may be not so straightforward to visualize the winding in this case, so it makes sense to delve a little more into this mapping. It is defined as

$$\theta(k) = \arg\{t_1 + t_2 \exp(\mathrm{i}\,k)\}\,. \tag{2.1.27}$$

Let us take a look at three extreme cases:

- $t_2 = 0, t_1 \neq 0\,,$
- $t_1 = 0, t_2 \neq 0\,,$
- $t_1 = t_2 \neq 0\,.$

In the first case, it is immediately clear that $\theta(k) = 0$ for all values of k. That is, $\theta(k)$ does not wind around the unit circle as k winds around the unit circle itself. Therefore, the winding number is $\nu = 0$ in that case. In the second case, the argument is directly k, so, similar to the Aharonov–Bohm case, a winding of k around S^1 leads to a winding around S^1 for $\theta(k)$. The winding number is therefore $\nu = 1$. In the third case, however, the winding number cannot be defined because a closed path for k around S^1 does not lead to a closed path around S^1 in this case, but rather to an open path. Indeed, in such a case $\theta(k) = k/2$. This has to do with the fact that the $t_1 = t_2$ line is that where the two bands touch at the edge of the Brillouin zone or, more precisely, it corresponds to the artificial folding of a one-dimensional chain with a single atom per unit cell. Therefore, that line, $t_1 = t_2$, corresponds to what is known as a *topological phase transition*. All other paths where $t_2 < t_1$ can be continuously connected to those where $t_2 = 0, t_1 \neq 0$, whereas all paths where $t_2 > t_1$ can be connected to those where $t_1 = 0, t_2 \neq 0$. The three cases are summarized in Fig. 2.2.

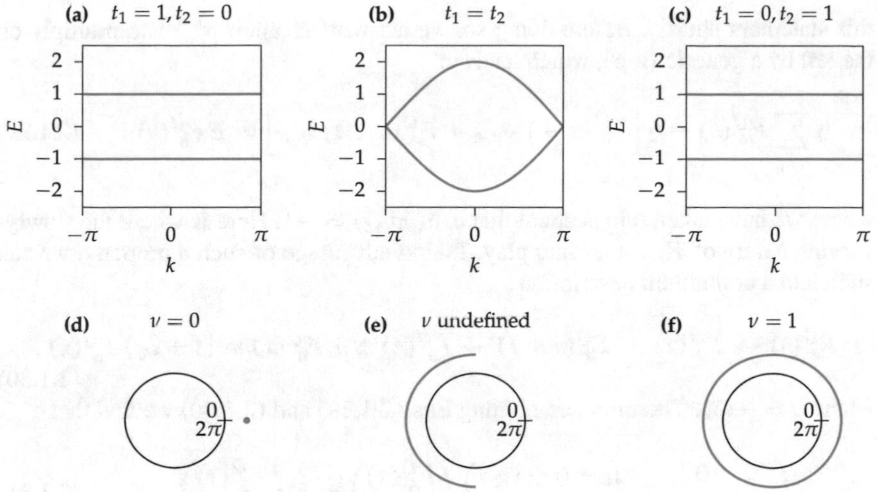

Fig. 2.2 Topology in the SSH model. Dispersion relations of the SSH chain for three extreme cases: **a** $t_1 = 1, t_2 = 0$, **b** $t_1 = t_2$ and **c** $t_1 = 0, t_2 = 1$. Mappings $\theta(k)$ from S^1 to S^1 with their corresponding winding numbers corresponding to the dispersion relation shown above each of these figures: **d** the loop is a single point and, therefore, has $\nu = 0$, **e** there is no closed loop so we cannot define a winding number and **f** the loop wraps around the circle, so $\nu = 1$. These maps have been enlarged in comparison to the black circles where they live for visualization purposes only

2.1.2 Zero Modes and the Bulk-Boundary Correspondence

In order to explore the consequences of having nonzero winding numbers, we will consider the *envelope function approximation* [5], which we shall also refer to as the *low energy continuum description* [18]. Except for the critical case, $t_1 = t_2$, the electronic dispersion displays two bands separated by an energy gap. As explained above, each carbon atom provides with one electron and, due to spin degeneracy of the band, the lower band is fully filled and the upper band is completely empty, leading to a regime known as *half filling*. We shall then be interested in those electrons closest to the Fermi energy, which in this case corresponds to those at the edge of the Brillouin zone, $Q = \pi$. Recall that distances are measured in units of the lattice spacing, a, so that crystal momenta are measured in units of $1/a$. With this in mind, let us rewrite Eq. (2.1.16) as follows

$$|\psi_k\rangle = \sum_\alpha \left(\sum_n e^{\mathrm{i}\, Qn} F_\alpha^Q(n)|n\rangle \right) \otimes |\alpha\rangle \, . \tag{2.1.28}$$

In order to coincide with Eq. (2.1.16), $F_\alpha^Q(n)$ would have wavevector $q = k - Q$. If we want to explore the vicinity of Q, then q is a small number and, therefore, $F_\alpha^Q(n)$ is a slowly-varying function of n around Q [19]. We shall see the consequences of this statement shortly. Before doing so, we act with H upon $|\psi_k\rangle$ and multiply on the left by a generic $\langle l, \beta|$, which leads to

$$t_1 \sum_{\alpha\neq\beta} F_\alpha^Q(n) - t_2 \left[F_B^Q(n-1)\delta_{\beta,A} + F_A^Q(n+1)\delta_{\beta,B} \right] = E F_\beta^Q(n) , \qquad (2.1.29)$$

where we have taken into account that $\exp(\pm \mathrm{i}\, Q) = -1$. Here is where the slowly-varying nature of F_α comes into play. Taking advantage of such a property, we can switch to a continuum description

$$F_\alpha^Q(n) \to F_\alpha^Q(x) , \quad F_\alpha^Q(n \pm 1) \to F_\alpha^Q(x) \pm \partial_x F_\alpha^Q(x) \equiv (1 \pm \mathrm{i}\, q)\, F_\alpha^Q(x) , \qquad (2.1.30)$$

where $q = -i\partial_x$. Therefore, combining Eqs. (2.1.29) and (2.1.30) we find that

$$\begin{pmatrix} 0 & t_1 - t_2 + \mathrm{i}\, t_2 q \\ t_1 - t_2 - \mathrm{i}\, t_2 q & 0 \end{pmatrix} \begin{pmatrix} F_A^Q(x) \\ F_B^Q(x) \end{pmatrix} = E \begin{pmatrix} F_A^Q(x) \\ F_B^Q(x) \end{pmatrix} . \qquad (2.1.31)$$

Notice that this equation is exactly the same as that for the amplitudes, $\varphi_\alpha(k)$, given in Eq. (2.1.23) when doing $k = q + Q$ and expanding around $q = 0$. This is the usual approach to obtain such a low energy Hamiltonian in a faster way from the microscopic, tight-binding description [18]. In order to keep the notation clearer and to study the effect of a small distortion, we shall denote $t_1 = t + \delta t$ and $t_2 = t - \delta t$, where $4|\delta t| \ll t$ is the bulk energy gap. Notice that δt can be either positive or negative. If negative, it corresponds to a nonzero bulk winding number, as discussed earlier in the text, and otherwise if positive. Let us denote by $m = 2\delta t$ and $v_F = t$. Then, we can write the previous equation as follows

$$\left(m\sigma_x - v_F\sigma_y q\right) \boldsymbol{F}^Q(x) = E\, \boldsymbol{F}^Q(x) , \qquad \boldsymbol{F}^Q(x) = \begin{pmatrix} F_A^Q(x) \\ F_B^Q(x) \end{pmatrix} . \qquad (2.1.32)$$

In order to make this equation more familiar, we can perform a rotation such that $\sigma_x \to \sigma_y$ and $\sigma_y \to -\sigma_x$. We do so with $\exp(-\mathrm{i}\,\pi\,\sigma_z/4)$. Hence,

$$\left(v_F\sigma_x q + m\sigma_y\right) \boldsymbol{F}^Q(x) = E\, \boldsymbol{F}^Q(x) . \qquad (2.1.33)$$

For clarity, we have chosen not to rename the rotated $\boldsymbol{F}^Q$. This is a Dirac equation in $1+1$ dimensions, where the speed of light is replaced by the *Fermi velocity* and the mass is replaced by the bulk energy gap. We will use this model to explore the consequences of the topology in the SSH chain. Since the model has two equivalent ground states, that is, one where $\delta t > 0$ and another where $\delta t < 0$ (both of same magnitude $|\delta t|$), one would expect the appearance of *domain walls*. That is, a region where the hopping sequence inverts. For instance, $\ldots, t_1 t_2 t_1 t_2 \mathbf{t_1 t_1} t_2 t_1 \ldots$. See Fig. 2.3.

In essence, it is a region where m changes from being $2\delta t$ to $-2\delta t$. Assuming $\delta t > 0$, this means that the bulk region with $m = 2\delta t$ corresponds to a sector with

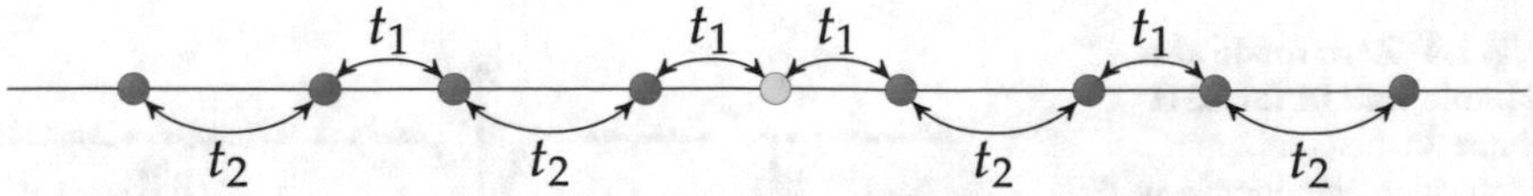

Fig. 2.3 **Domain wall in the SSH chain**. A chain with ordering t_2t_1 abruptly changes to t_1t_2. The yellow spot corresponds to the location of the domain wall and it will be a place for finding a zero mode

winding number $\nu = 0$ and the bulk region with $m = -2\delta t$ corresponds to a sector with winding number $\nu = 1$. Since the domain wall connects two regions of different winding numbers, we say that it is a *topological boundary* [20]. The consequence, which translates to many other examples of topological phases with invariants different from the winding number [21–23], will be the presence of topologically protected states at the boundary. This is known as the *bulk-boundary correspondence* [16, 23–25]. Let us explicitly derive this by introducing a mass that changes sign in the Dirac equation. Although this could be done smoothly, for our purposes we can assume the interface to be abrupt and propose $m = |m|\text{sgn}\,(x)$. As is customary with the Dirac equation, in order to find the spectrum we study the squared Hamiltonian

$$\left[-v_F^2\partial_x^2 + m^2 + 2v_F|m|\sigma_z\delta(x)\right] \boldsymbol{F}^Q(x) = E^2\,\boldsymbol{F}^Q(x)\,. \tag{2.1.34}$$

The solution to this equation is given by

$$\boldsymbol{F}^Q(x) = \exp\left(-k|x|\right)\boldsymbol{\Phi}\,, \qquad k^2 = \frac{m^2 - E^2}{v_F^2}\,, \tag{2.1.35}$$

where $\boldsymbol{\Phi}$ is a constant vector which, due to continuity, it is the same on both sides of the domain wall, and we have taken $k > 0$. Notice that this solution implies that the energy of the topological boundary mode has to be within the bulk energy gap. Integrating Eq. (2.1.34) we can obtain the other boundary condition

$$-\,v_F^2\left[\partial_x\boldsymbol{F}(0^+) - \partial_x\boldsymbol{F}(0^-)\right] + 2v_F|m|\sigma_z\boldsymbol{F}(0) = 0\,, \tag{2.1.36}$$

or, equivalently,

$$\sigma_z\boldsymbol{\Phi} = -\frac{kv_F}{|m|}\boldsymbol{\Phi}\,. \tag{2.1.37}$$

This can be thought of as an eigenvalue problem and we can take as eigenvectors those of σ_z. However, there is also the restriction that $k > 0$, which only allows us to consider the eigenvalue -1 with eigenvector $\boldsymbol{\Phi} = (0, 1)^T$. Therefore, we find that $kv_F = |m|$ which, taking into account Eq. (2.1.35), it implies the existence of a non-degenerate, zero-energy mode, $E = 0$, localized at the interface

$$\boldsymbol{F}^Q(x) = \exp\left(-\frac{|mx|}{v_F}\right)\begin{pmatrix}0\\1\end{pmatrix}\,, \qquad E = 0\,. \tag{2.1.38}$$

Fig. 2.4 Zero mode at a domain wall in the SSH chain. In the continuum description, the domain wall amounts to a change in the mass term of the Dirac equation

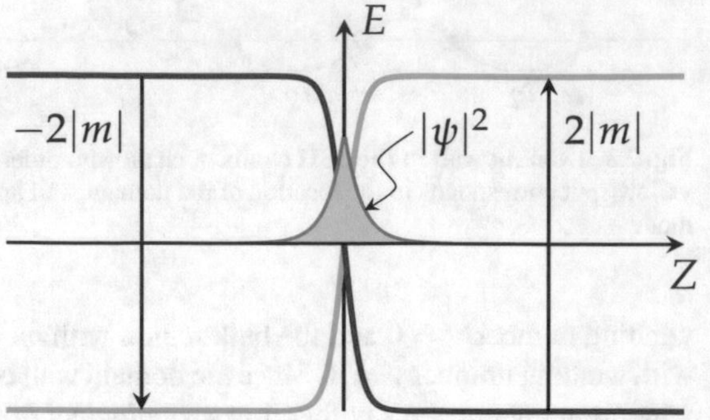

This zero-energy mode was obtained by SSH [13], and a similar result from a seminal paper by Jackiw and Rebbi was obtained in the context of field theory [26]. We show the zero mode schematically in Fig. 2.4.

2.1.3 Symmetries and the Altland–Zirnbauer Classification

Up to now, we have only discussed topology and two particular setups where one can define winding numbers. However, the title of this section also contains the word *symmetry*. This is particularly relevant to our last example. These symmetries are present in the microscopic model and transfer directly to the continuum model as well.[2] The following discussion will prove to be useful also for constructing low-energy Hamiltonians with the *method of invariants* [10]. The argument is as follows: let $\mathcal{H}(\mathcal{K})$ be a Hamiltonian of a system depending on a general tensor operator, $\mathcal{K}$, which may depend on the crystal momentum, electric and magnetic fields, strain, and the spin of the electrons [27]. Then, if the system is invariant under an element g of a symmetry group G, then the Hamiltonian must also be invariant under such a symmetry, that is [10]

$$\mathcal{D}(g)\mathcal{H}(\mathcal{K})\mathcal{D}^{-1}(g) = \mathcal{H}(g\mathcal{K}) \,, \tag{2.1.39}$$

where $\mathcal{D}(g)$ is the matrix representation of the group element g. A theorem in group theory states that we can always choose such a matrix representation to be unitary [28]. It is also important to consider time-reversal invariance, $\mathcal{T}$. In that case, we have

$$\Theta\mathcal{H}(\mathcal{K})\Theta^{-1} = \mathcal{H}(\zeta\mathcal{K}) \,, \tag{2.1.40}$$

where $\zeta = +1$ for strain and electric field and $\zeta = -1$ for crystal momentum, spin and magnetic field. That is, under time-reversal symmetry, the strain tensor or the

[2]The continuum model may display more symmetries than the microscopic model does due to restricting terms to lowest order in momenta. We shall see examples of this throughout the text.

electric field are unaffected, whereas momentum is inverted (thus inverting the group velocity) and also are the spin and the magnetic field, because these two are pseudovectors. Here, Θ is the operator representing the operation of $\mathcal{T}$. However, in contrast to the matrix representations $\mathcal{D}(g)$, Θ cannot be chosen to be unitary. Although the argument can be made rigorous [17], we shall keep it simple: time-reversal symmetry does not affect position, but it changes the sign of momentum. Consider the canonical commutator

$$\Theta\,[x, p]\,|\psi\rangle = \Theta \mathrm{i}\,|\psi\rangle\,, \tag{2.1.41}$$

where $|\psi\rangle$ is an arbitrary state. Then, if Θ^{-1} exists, we can insert the identity and write

$$\Theta\,[x, p]\,\Theta^{-1}\Theta|\psi\rangle = \Theta \mathrm{i}\,|\psi\rangle\,. \tag{2.1.42}$$

As we said earlier, time reversal inverts momentum but not position, so $\Theta\,[x, p]\,\Theta^{-1} = [x, -p] = -\,[x, p]$. Hence,

$$-\,[x, p]\,\Theta|\psi\rangle = \Theta \mathrm{i}\,|\psi\rangle\,. \tag{2.1.43}$$

If Θ were linear, then $\Theta c|\psi\rangle = c\Theta|\psi\rangle$, where c is a complex number. This in turn would not leave the commutator invariant. Instead, we require Θ to be antilinear, so that $\Theta c|\psi\rangle = c^{*}\Theta|\psi\rangle$. As a result,

$$[x, p]\,\Theta|\psi\rangle = \mathrm{i}\,\Theta|\psi\rangle\,. \tag{2.1.44}$$

Therefore, $[x, p]$ acting upon the time-reversed, arbitrary state $\Theta|\psi\rangle$, still leads to the same i. Therefore, if Θ is antilinear and Θ^{-1} exists, the canonical commutator stays invariant under time-reversal symmetry. Moreover, Θ has to be such that the norm of $|\psi\rangle$ is invariant under its action. An operator satisfying these three conditions is said to be *antiunitary* [6, 17]. The actual form of Θ will depend on the representation, but we may generally write it as the product of a unitary operator, U, and complex conjugation, $\mathcal{K}$, which would satisfy all the requirements stated before. Hence, we write

$$\Theta = U\mathcal{K}\,. \tag{2.1.45}$$

Two particular realizations of U are of importance. One is that for spinless particles and the other is that for spinful particles. For simplicity, we shall restrict the following discussion to the case where the orbital variables are given in the coordinate representation and spin will be in the representation where S_z is diagonal. Other representations (e.g. the momentum representation), can be considered for the orbital variables [17], but the coordinate representation is particularly useful because the basis states are unaffected by Θ. In such a case, the unitary operator is therefore just the identity. For particles with spin, however, spin has to change sign under time-reversal symmetry, as discussed above. Therefore, complex conjugation alone only changes the sign of S_y, which is imaginary in the chosen representation for the spin variables. Hence, we need a unitary operator that changes the sign of S_x and

S_z, while leaving S_y untouched. Also, the unitary operator must only act on the spin subspace, so that it does not affect the orbital variables, which are already taken care of by complex conjugation. We can achieve such a result by means of a π rotation in the spin space around the S_y axis. That is, U must be given by [17]

$$U = \exp\left(-\mathrm{i}\pi S_y\right) . \tag{2.1.46}$$

In the case of spin-1/2 particles, where $S_y = \sigma_y/2$, using the property that $\sigma_y^2 = \mathbb{1}_2$, it is particularly simple to show that

$$U = -\mathrm{i}\sigma_y . \tag{2.1.47}$$

It is also particularly interesting to study the square of Θ, since one would normally argue that applying Θ twice should be equivalent to the identity. Surprisingly, this is not the case. Indeed, since i S_y is real, then U and $\mathcal{K}$ commute and we find that

$$\Theta^2 = \exp\left(-\mathrm{i}\,2\pi S_y\right) . \tag{2.1.48}$$

As we can see, Θ^2 is unitary and its eigenvalues are either $+1$ or -1 depending on whether we have integer or half-integer spins, respectively. There are important consequences to these that we will explore in due time. As of now, let us come back to the SSH system. Let us first start with the more conventional symmetry groups. In this case, the only symmetry we find is space inversion,[3] I, about the center of an arbitrary bond center (remember we are considering the infinite chain). This symmetry takes A to B and vice versa within a unit cell, so in order to hold A and B must actually be chemically equivalent atoms, as is the case because A and B are both carbon atoms. Hence, a unitary matrix representation of inversion that exchanges A and B can be taken to be $\mathcal{D}(I) = \sigma_x$. We also need to know the action of I on the general tensor, $\mathcal{K}$. In this case, we are neglecting strain, electric and magnetic fields, and spin. Therefore, we are only left with the crystal momentum, k, which changes sign under inversion. Hence, space inversion is finally written as follows

$$\sigma_x \mathcal{H}(k) \sigma_x = \mathcal{H}(-k) . \tag{2.1.49}$$

Let us now turn our attention to time-reversal symmetry. Since we are considering spinless particles or, equivalently, spin-polarized, then $\Theta = \mathcal{K}$. Hence, taking into account that k changes sign under time-reversal, we can write

$$\mathcal{H}^*(k) = \mathcal{H}(-k) . \tag{2.1.50}$$

Consequently, the two symmetries would hold if

[3] We have already exploited translational symmetry to block-diagonalize the tight-binding Hamiltonian. That is, we have block-diagonalized the Hamiltonian according to the irreducible (one-dimensional) representations of the translation operator.

$$\sigma_x \mathcal{H}^*(k)\sigma_x = \mathcal{H}(k) \,. \tag{2.1.51}$$

If we write a generic two-band Hamiltonian as follows,

$$\mathcal{H}(k) = \boldsymbol{d}(k) \cdot \boldsymbol{\sigma} \,, \tag{2.1.52}$$

we can see that the combination of the two symmetries leads to $d_z(k) = 0$. On the other hand, separately these symmetries lead to $d_x(k) = d_x(-k)$ and $d_y(k) = -d_y(-k)$. In the microscopic model of the SSH chain given by Eq. (2.1.23), the $\boldsymbol{d}(k)$ vector is given by

$$\boldsymbol{d}(k) = (\mathrm{Re}\,[\delta(k)]\,, \mathrm{Im}\,[\delta(k)]\,, 0) = (t_1 + t_2\cos(k), t_2\sin(k), 0) \,. \tag{2.1.53}$$

Notice that, indeed, the two symmetries that we have discussed hold in this scenario. This system has two more symmetries that are somehow more subtle, but it is their presence that leads to the peculiar topological behaviour of the SSH model. The first symmetry is a unitary operator, but it does not behave as a usual unitary, in the sense that it does not commute with the Hamiltonian. Rather, it anticommutes. This is known as *chiral* or *sublattice symmetry*, Γ, which is such that

$$\Gamma\mathcal{H}(k)\Gamma = -\mathcal{H}(k) \,. \tag{2.1.54}$$

Before we apply this to the SSH model, let us pause and ponder on the implications of this symmetry. Let $|\psi\rangle$ be an eigenstate of the Hamiltonian, with energy E. Then, due to the anticommutation of the chiral operator with the Hamiltonian, there will be another eigenstate $\Gamma|\psi\rangle$ to the Hamiltonian with energy $-E$, enforcing the spectrum to be symmetric around $E = 0$. However, it is possible that $|\psi\rangle$ turns out to be an eigenstate of Γ, in which case $|\psi\rangle$ and $\Gamma|\psi\rangle$ are actually the same state. This seems to contradict our discussion that chiral symmetry enforces states to come in pairs $(E, -E)$. Therefore, the only option for a state to be its own chiral partner is to be at zero energy, that is, to be a zero-energy mode. As a result, as long as there are no chiral-symmetry-breaking perturbations, the zero-energy mode will remain at zero energy. In the SSH model, we can see that Γ is realized in σ_z. Therefore, as long as there are no diagonal terms in the Hamiltonian, chiral symmetry will be preserved. Notice that constant diagonal terms proportional to the identity are valid, since those can be absorbed in the energy. However, if there are terms proportional to the identity but depending on k, they cannot be absorbed in a redefinition of the energy. Those terms break chiral symmetry, and they would naturally appear when including second-nearest-neighbour hoppings. Let us continue our discussion ignoring such terms. We said that there were two more symmetries apart from space and time reversal, one of them being chiral symmetry. The other symmetry is an antiunitary operator that is obtained as the product of time-reversal symmetry and chiral symmetry and it is commonly referred to as *particle-hole* or *charge conjugation symmetry*. Therefore,

$$\mathcal{P}\mathcal{H}(\mathcal{K})\mathcal{P}^{-1} = -\mathcal{H}(\zeta\mathcal{K}) \,, \tag{2.1.55}$$

Table 2.1 Altland–Zirnbauer (AZ) classification of topological matter. Single particle Hamiltonians are categorized depending on the presence or absence of time-reversal, $\mathcal{T}$, particle-hole $\mathcal{P}$ and chiral, $\mathcal{C}$, symmetries. Absence of a symmetry is indicated with a 0. Presence of $\mathcal{T}$ or $\mathcal{P}$ is labeled with ± 1, depending on the value of the corresponding squared operators. Presence of chiral symmetry is indicated by 1. Depending on spatial dimension $(1, 2, 3)$, the space of ground states may be trivial, as indicated by a dash. It may be divisible into a countably infinite number of topological sectors, each of which is labeled by an integer $\mathbb{Z}$. Finally, there may be only two distinct topological sectors, allowing us to label them with a binary or $\mathbb{Z}_2$ quantity [33]

AZ class	Symmetry			Dimension		
	$\mathcal{T}$	$\mathcal{P}$	$\mathcal{C}$	1	2	3
A	0	0	0	–	$\mathbb{Z}$	–
AI	+1	0	0	–	–	–
AII	−1	0	0	–	$\mathbb{Z}_2$	$\mathbb{Z}_2$
AIII	0	0	1	$\mathbb{Z}$	–	$\mathbb{Z}$
BDI	+1	+1	1	$\mathbb{Z}$	–	–
CII	−1	−1	1	$\mathbb{Z}$	–	$\mathbb{Z}_2$
D	0	+1	0	$\mathbb{Z}_2$	$\mathbb{Z}$	–
C	0	−1	0	–	$\mathbb{Z}$	–
DIII	−1	+1	1	$\mathbb{Z}_2$	$\mathbb{Z}_2$	$\mathbb{Z}$
CI	+1	−1	1	–	–	$\mathbb{Z}$

where $\mathcal{P} = \Gamma\Theta$. Just like with time-reversal symmetry, $\mathcal{P}$ can also square to ± 1 [29]. In our case $\mathcal{P}^2 = 1$. It is important to notice that all these symmetries that we have discussed are also present in the continuum Hamiltonian.

With all this knowledge, we can try to find a place for the SSH model in the Altland–Zirnbauer classification [29–33]. That classification does something similar to our discussion of homotopy. However, it is rather involved and we shall restrict to say that, in that classification, the homotopy groups of Hamiltonians are given according to the spatial dimension and the symmetries. As a result, each ground state of the Hamiltonian is labeled with a topological invariant. Hence, different gapped ground states of different topological invariants cannot be connected because they belong to different sectors. In total, there are ten symmetry classes resulting from the different combinations of having or not time-reversal, particle-hole and chiral symmetries and whether the former two square to $+1$ or -1. A table with all such symmetries and the corresponding homotopy groups for spatial dimensions $d = 1, 2, 3$ is shown in Table 2.1 [30, 33]. Whenever the symbol $\mathbb{Z}$ appears it implies that there is a countably infinite number of topological sectors, each of which can be labeled by an integer. If $\mathbb{Z}_2$ appears instead, there are only two distinct topological sectors that can be labeled with a binary quantity. In our case, we have time-reversal and particle-hole symmetries, both squaring to $+1$, and chiral symmetry. Therefore, we can place the SSH model in the BDI class. In that case, the space of ground states is partitioned into sectors with different $\mathbb{Z}$ invariant. As we have obtained, the phases with $t_1 > t_2$ and $t_1 < t_2$ are both ground states of the Hamiltonian with the same energy gap of $2|t_1 - t_2|$, but they are topologically distinct, since the former has winding number $\nu = 0$ and the latter has $\nu = 1$. That is, it is not possible to continuously connect both ground states without undergoing a topological phase transition at $t_1 = t_2$. One can access other ground states with higher winding numbers in the BDI class by allowing only odd-neighbour hoppings [34]. In that case, all symmetries are preserved and we are still in the BDI class. Notice that this is far from realistic, since one would expect that in reality there should also be even-neighbour hoppings, which break both chiral and particle-hole symmetries, placing the SSH model in the AI class, which is trivial in one dimension. In this case, all gapped ground states are topologically equivalent and one can continuously shift from one to the other without closing the gap. Indeed, if we allow for a constant term $d_z(k) = \lambda$, the spectrum would become

$$E(k) = \pm\sqrt{t_1^2 + t_2^2 + 2t_1t_2\cos(k) + \lambda^2}\,, \tag{2.1.56}$$

which at $k = \pi$ would lead to

$$E(Q) = \pm\sqrt{(t_1 - t_2)^2 + \lambda^2}\,. \tag{2.1.57}$$

Hence, if $\lambda \neq 0$, we can shift from $t_1 > t_2$ to $t_1 < t_2$ without closing the gap and, as a result, the two ground states are topologically equivalent. If we create a domain wall and proceed as before, the state at zero energy moves to $E = \lambda$. Since chiral symmetry is absent, there is no need for an $E = -\lambda$ state to stabilize the former. Assuming $\lambda > 0$, as it increases keeping m fixed, the state moves higher up in energy until it reaches the continuum of states and becomes delocalized. This very simple model of a perturbation of the form $\lambda\sigma_z$ is known as the Rice–Mele model [35, 36]. By solving it on a finite system, it is possible to observe the aforementioned disappearance of the zero modes into the bulk.

2.2 Method of Invariants

The method of invariants allows us to obtain low-energy continuum Hamiltonians as the one we obtained for the SSH model by taking into account the symmetries of the system. A detailed account of this method can be found in Refs. [7–10]. Although it is a truly powerful method for obtaining Hamiltonians for arbitrary tensor operators $\mathcal{K}$, we shall restrict ourselves to the case where $\mathcal{K}$ contains only the crystal momentum, $\boldsymbol{k}$, and spin. Using group-theoretical arguments, one can obtain the symmetry-allowed terms of the Hamiltonian by means of what is known as an *invariant expansion*. In order to apply the method, we first need to choose a basis and explore how does it transform under the symmetries of the system. For the SSH model, we choose as a basis $\{\psi_A(x), \psi_B(x)\}$. The system possesses inversion symmetry, I, which in this representation amounts to

$$\psi_A(x) \to \psi_A(-x) = \psi_B(x) , \quad (2.2.1a)$$
$$\psi_B(x) \to \psi_B(-x) = \psi_A(x) . \quad (2.2.1b)$$

Therefore, the basis functions transform according to the representation $\mathcal{D}(I) = \sigma_x$. On the other hand, since the wavenumber k transforms under inversion as $k \to -k$, the invariance condition requires that [cf. (2.1.40)]

$$\sigma_x \mathcal{H}(k) \sigma_x = \mathcal{H}(-k) . \quad (2.2.2)$$

On the other hand, spinless time-reversal symmetry requires that

$$\mathcal{H}(k) = \mathcal{H}^*(-k) . \quad (2.2.3)$$

Let $h_{ij}(k)$ be the matrix elements of $\mathcal{H}(k)$ with $i, j = 1, 2$. Inversion symmetry leads to the following constraints to the matrix elements

$$h_{22}(k) = h_{11}(-k) , \qquad h_{21}(k) = h_{12}(-k) . \quad (2.2.4)$$

Time-reversal symmetry, on the other hand, requires that $h_{ij}(k) = h^*_{ij}(-k)$. Since the matrix has to be Hermitian as well, the diagonal coefficients have to be real and, therefore, even in k, that is, $h_{11}(k) = h_{11}(-k)$. Inversion symmetry also implies that the diagonal entries contribute to a term proportional to the identity, but not to σ_z. The off-diagonal entries are complex in general, but to ensure Hermiticity they are such that $h_{12}(k) = h^*_{21}(k)$. Time-reversal symmetry requires that $h_{12}(k) = h^*_{12}(-k)$. As a result, there are no restriction to the evenness or oddness as a function of k for $h_{12}(k)$, as long as such restriction is fulfilled. Therefore, with this knowledge we can expand the matrix coefficients as follows

$$h_{11}(k) \simeq a_{11} + b_{11}k^2 + \mathcal{O}(k^4) , \quad (2.2.5a)$$
$$h_{12}(k) \simeq a_{12} + b_{12}k + c_{12}k^2 + \mathcal{O}(k^3) , \quad (2.2.5b)$$

where the constant coefficients will be real in $h_{11}(k)$. In $h_{12}(k)$, however, they will be real for even powers of k and purely imaginary for odd powers of k. Restricting the expansion to the lowest powers in k, we find that $\mathcal{H}(k)$ is given by

$$\mathcal{H}(k) \simeq a_{11}\mathbb{1}_2 + a_{12}\sigma_x + |b_{12}|k\,\sigma_y . \quad (2.2.6)$$

Let us give the coefficients in the expansion an interpretation. In order to do so, we can diagonalize the Hamiltonian to obtain

$$E(k) = a_{11} \pm \sqrt{a_{12}^2 + |b_{12}|^2 k^2} . \quad (2.2.7)$$

If $k = 0$, we should obtain the two band edges, that is, the minimum and maximum of the conduction and valence bands, respectively. Therefore, $a_{11} + a_{12} = \varepsilon_c$ and

$a_{11} - a_{12} = \varepsilon_v$, where ε_c and ε_v are the band-edge energies. If ε_0 is the gap center and $2m$ is the energy gap, we can write $\varepsilon_c = \varepsilon_0 + m$ and $\varepsilon_v = \varepsilon_0 - m$. Hence, $a_{11} = \varepsilon_0$ and $a_{12} = m$. On the other hand, when $m = 0$ the system becomes metallic and we should be able to expand the true dispersion relation linearly around the Fermi energy. That is, $E(k) \simeq \varepsilon_0 \pm v_F k$. Hence, $|b_{12}| = v_F$. Finally, we can write the Hamiltonian as follows

$$\mathcal{H}(k) = \varepsilon_0 \mathbb{1}_2 + m\sigma_x + v_F k\, \sigma_y\ , \tag{2.2.8}$$

which is exactly the Hamiltonian we obtained for the SSH chain by means of the $\boldsymbol{k} \cdot \boldsymbol{p}$ approximation, now using only symmetry arguments. The only difference between the two is the constant term $\varepsilon_0 \mathbb{1}_2$, which can be absorbed in the energy.

We will now proceed to find the low-energy Hamiltonian for graphene and for the three-dimensional topological insulators that will be the subject of this thesis.

2.2.1 *Graphene Low-Energy Hamiltonian*

Let us start with graphene (see Ref. [37] for a review on the electronic properties of graphene). The formation of the graphene lattice is very similar to that of polyacetylene that we discussed earlier for the SSH model. The only difference is that, instead of placing hydrogen atoms in the unpaired sp^2 orbital, we place another sp^2 orbital. The result is a honeycomb lattice or, rather, a triangular Bravais lattice with a double basis. We shall denote the atoms in the basis as A and B, respectively. Just like in the SSH chain, the p_z orbitals weakly couple forming π bonds and, as a consequence of the large number of orbitals involved in the bulk crystal, these form a π and a π^* band. There are also σ bands from the σ bonds between the sp^2 orbitals, but these are far below or above the Fermi energy and we shall not be interested in those. As of now, the system is very similar to the SSH model, in the sense that we have these two π and π^* bands coming from the out-of-plane p_z orbitals belonging to the two sublattices. In fact, since each carbon atom carries one electron on each p_z orbital, the π band is fully occupied and the π^* band is empty. However, there are important differences between the two systems, as we will see shortly. For the following discussion it is convenient to have Fig. 2.5 in mind. The lattice vectors, shown in Fig. 2.5a, in units of the lattice constant a are given by

$$\boldsymbol{a}_1 = \frac{1}{2}\left(\sqrt{3}, 1\right)\ , \qquad \boldsymbol{a}_2 = (0, 1)\ . \tag{2.2.9}$$

The positions of the A atoms will be linear combinations of these two vectors with integer coefficients, that is,

$$\boldsymbol{R}_A = m\boldsymbol{a}_1 + n\boldsymbol{a}_2\ . \tag{2.2.10}$$

The B atoms are connected to the A atoms by a vector $\boldsymbol{\delta} = (-1/\sqrt{3}, 0)$, so that the positions of the B atoms are $\boldsymbol{R}_B = \boldsymbol{R}_A + \boldsymbol{\delta}$. The resulting honeycomb lattice is

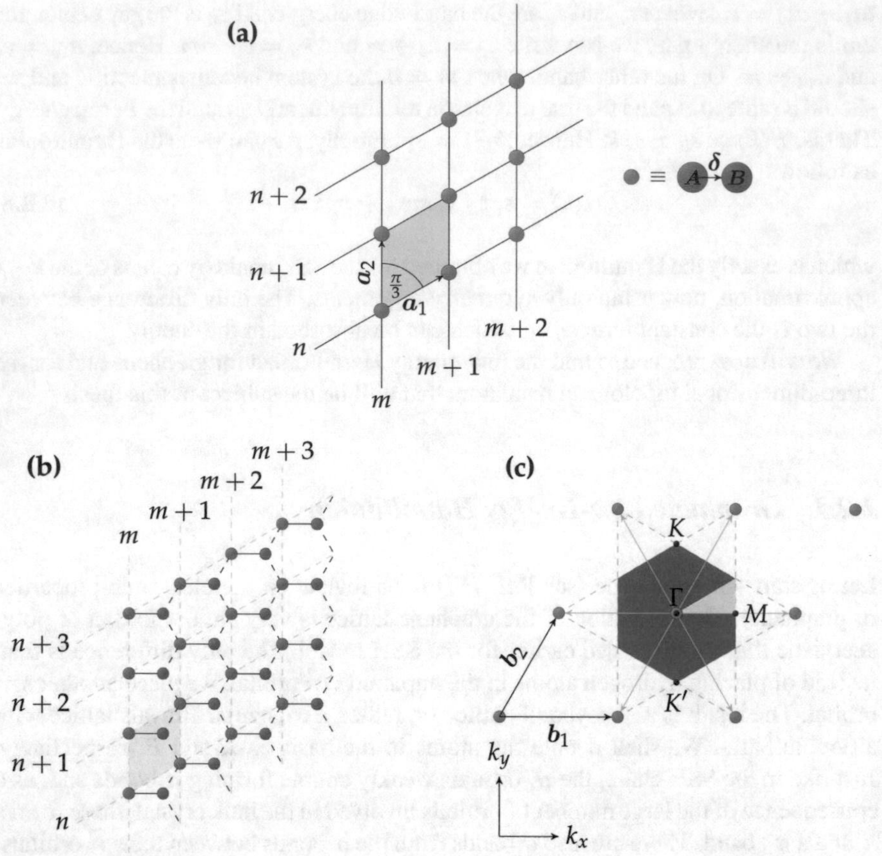

Fig. 2.5 Crystallography of the honeycomb structure. **a** Triangular lattice underlying the honeycomb structure. The lattice vectors, $\boldsymbol{a}_1$ and $\boldsymbol{a}_2$ are shown, together with a choice for the unit cell (shaded rhombus). On the right-hand side the basis formed by A and B is shown, together with the vector $\boldsymbol{\delta}$ that links the two. **b** Honeycomb lattice, resulting from placing the basis of A and B atoms on each of the points of the triangular lattice. Solid lines have been used to emphasize the basis, whereas black dashed lines are shown to complete the hexagons. **c** Reciprocal lattice (grey dots) spanned by the reciprocal lattice vectors, $\boldsymbol{b}_1$ and $\boldsymbol{b}_2$, are shown along with the hexagonal Brillouin zone (blue-shaded area) and the high-symmetry points, Γ, M, K, K'. The irreducible Brillouin zone is shown in yellow

shown in Fig. 2.5b. The reciprocal lattice vectors are easily obtained by the standard procedures and we find

$$\boldsymbol{b}_1 = \frac{2\pi}{\sqrt{3}}(2, 0) , \qquad \boldsymbol{b}_2 = \frac{2\pi}{\sqrt{3}}(1, \sqrt{3}) . \tag{2.2.11}$$

Linear combinations of these two with integer coefficients create the reciprocal lattice. Since we are in two-dimensions, the reciprocal lattice is obtained easily from the direct lattice by simply rotating $\pi/2$ and scaling appropriately. Therefore, the reciprocal lattice is also a triangular lattice and, as a result, the Brillouin zone is a hexagon rotated $\pi/2$ with respect to the original hexagonal Wigner–Seitz cell. The high-symmetry points of the Brillouin zone are the Γ, M, K and K' points, and the triangular wedge $\Gamma M K \Gamma$ defines the irreducible Brillouin zone. The reciprocal lattice, along with the Brillouin zone and the high-symmetry points is shown in Fig. 2.5c. Of special importance to us are the K and K' points, which correspond to two consecutive corners of the Brillouin zone. Those two points cannot be joined by a reciprocal lattice vector $\boldsymbol{G} = h\boldsymbol{b}_1 + l\boldsymbol{b}_2$, just like the position of a B atom cannot be reached by means of the direct lattice vectors. Taking into account the geometry of the hexagon, we can also choose for the K and K' points to be located in opposite sides of the hexagon

$$\boldsymbol{K} = -\boldsymbol{K}' = \frac{4\pi}{3}(0, 1) \,. \tag{2.2.12}$$

A very simple nearest-neighbour tight-binding model can be proposed for graphene working in the basis $|\boldsymbol{R}_\alpha\rangle \equiv |m, n\rangle \otimes |\alpha\rangle$

$$H = -t\left[\mathbb{1}_N \otimes \sigma_x + \left(T_x + T_y\right) \otimes \sigma_+ + \left(T_{-x} + T_{-y}\right) \otimes \sigma_-\right] \,, \tag{2.2.13}$$

where we have introduced the translation operators in the lattice subspace

$$T_{\pm x} = \sum_{m,n} |m \pm 1, n\rangle\langle m, n| \,, \qquad T_{\pm y} = \sum_{m,n} |m, n \pm 1\rangle\langle m, n| \,, \tag{2.2.14}$$

and $\sigma_\pm$ are the raising and lowering operators in the cell subspace. Proceeding in the same manner as with he SSH model we obtain a very similar result

$$\begin{pmatrix} 0 & \Delta^*(\boldsymbol{k}) \\ \Delta(\boldsymbol{k}) & 0 \end{pmatrix} \begin{pmatrix} \varphi_A \\ \varphi_B \end{pmatrix} = \frac{E(\boldsymbol{k})}{t} \begin{pmatrix} \varphi_A \\ \varphi_B \end{pmatrix} \,, \tag{2.2.15}$$

where $\boldsymbol{k} = (k_x, k_y)$ and

$$\Delta(\boldsymbol{k}) = 1 + \exp\left(\mathrm{i}\,\boldsymbol{k} \cdot \boldsymbol{a}_1\right) + \exp\left(\mathrm{i}\,\boldsymbol{k} \cdot \boldsymbol{a}_2\right) \,. \tag{2.2.16}$$

The dispersion is therefore given by

$$E(k) = \pm t|\Delta(\boldsymbol{k})| \,, \tag{2.2.17}$$

and it is shown in Fig. 2.6.

In the SSH model, the spectrum is gapped unless $t_1 = t_2$, which is a situation that would never occur because it is unstable. However, in graphene, the undistorted honeycomb lattice is the most stable one, which is why we have set all nearest-neighbour hoppings to be equal to t. However, in contrast to the SSH model, we can find a value of $\boldsymbol{k}$ where the two bands touch, that is, where $|\Delta(\boldsymbol{k})| = 0$. The reason is

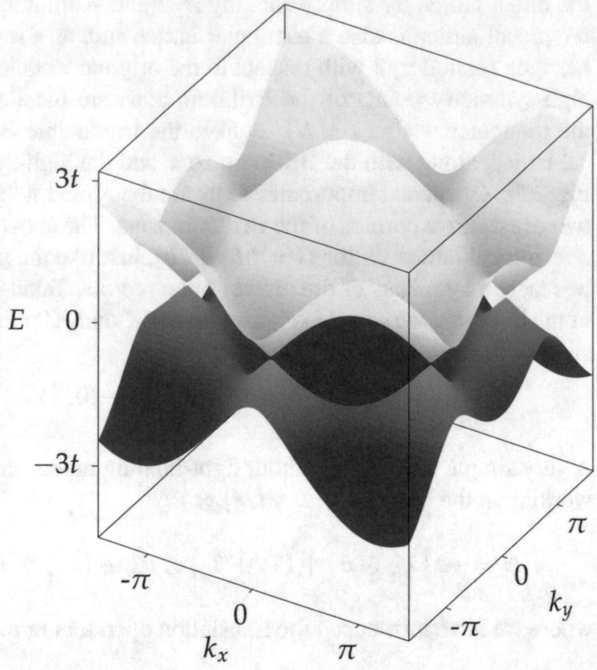

Fig. 2.6 Graphene dispersion relation. The two π-bands touch at the K and K' points of the Brillouin zone and the spectrum is linear around those points

that we have two equations for this to happen, $\mathrm{Re}[\Delta(\boldsymbol{k})] = 0$ and $\mathrm{Im}[\Delta(\boldsymbol{k})] = 0$, with two unknowns, k_x and k_y, so there is a possibility to find such degeneracies. The fact that there are no terms in the diagonal of the Hamiltonian allows us to circumvent the Wigner–von-Neumann theorem, which states that accidental degeneracies can only be achieved by fine-tuning of three parameters [38, 39]. In fact, the degeneracy of the SSH model when $t_1 = t_2$ is accidental, in the sense that it is not due to any symmetries. Indeed, that degeneracy occurs at $k = \pi$, when the hoppings are identical and the elements in the diagonal are set to zero. In graphene, the degeneracies where $|\Delta(\boldsymbol{k})| = 0$ come in pairs and appear at the $\boldsymbol{K}$ and $\boldsymbol{K}'$ points or *valleys*, respectively, which we shall call *Dirac points* for reasons that will become clear shortly. One could then argue that this result is in fact accidental, because it is peculiar to the honeycomb lattice. However, this is not the case. Indeed, one could set a third-neighbour hopping term, t', such that one can study the cases where $t' = 0$ (honeycomb lattice) up to $t' = t$ (square lattice). It can be shown that in that range of t' the degeneracies persist, meaning that the degeneracies must result from a symmetry and not from the fact that we have a honeycomb lattice [40, 41]. The symmetry involved is chiral symmetry like in the SSH chain, since there is a unitary operator Γ such that the Hamiltonian anticommutes with such an operator and it is also such that $\Gamma^2 = \mathbb{1}_2$. This operator is realized here in σ_z and it implies that the $\boldsymbol{d}(\boldsymbol{k})$ vector of the two-band Hamiltonian is restricted to a plane. This key fact allows us to understand the topological protection of the Dirac points and why they must appear in pairs. The argument is as follows: in general, $\boldsymbol{d}(\boldsymbol{k})$ is a mapping from the Brillouin zone torus into three-dimensional

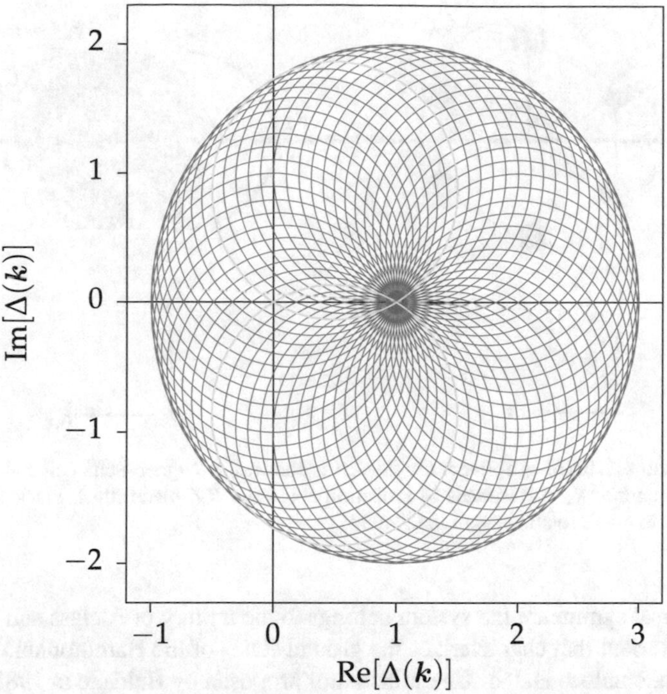

Fig. 2.7 Topology of bulk graphene. $\Delta(\boldsymbol{k})$ in the complex plane as $\boldsymbol{k}$ is taken through the Brillouin zone. In the figure, we have discretized k_y and let k_x vary. In truth, plotting for all values of k_y leads to a flat surface, as described in the text. The yellow circles are the only ones that touch the origin, which do so at the K and K' points [42]

space. Since the torus is periodic, the tip of $\boldsymbol{d}(\boldsymbol{k})$ traces out the image of a compact, oriented surface. Chiral symmetry then flattens that surface and, if it encloses the origin, then there will be two points of the original surface which come through the origin [42]. We can see this by plotting $\Delta(\boldsymbol{k})$ on the complex plane as $\boldsymbol{k}$ is evolved through the Brillouin zone. The result is shown in Fig. 2.7. Having $\boldsymbol{k}$ to fully traverse the Brillouin zone leads to a flat surface, as expected. The yellow circles are the only ones that touch the origin and correspond to setting $k_y = \pm 4\pi/3$ and letting k_x vary, the origin being touched when $k_x = 0$. That is, the intersection between those two circles and the origin correspond to the Dirac points.

The only way to get rid of the Dirac points is by merging them, which occurs at the boundary of the flattened surface. After merging, the spectrum becomes gapped. We can also gap the spectrum by breaking chiral symmetry. Doing so, we can then try to find the topological classification of such a system in the Altland–Zirnbauer classification (see Table 2.1) [29–33]. If we consider spinless fermions and preserve time-reversal symmetry, then the system belongs to the so-called orthogonal or AI class and it is topologically trivial. An example of this is hexagonal boron nitride, where the atoms in the two sublattices are chemically different [43]. If we break

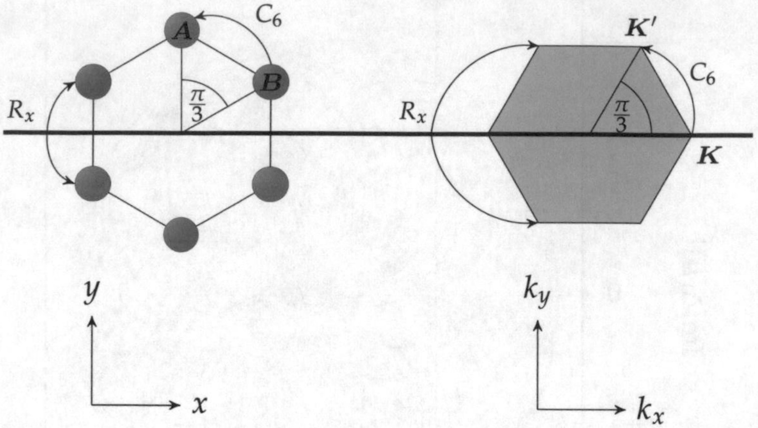

Fig. 2.8 **Symmetries of graphene**. On the left is shown the Wigner–Seitz cell and on the right the Brillouin zone. R_x corresponds to reflection about the XZ plane (thick black line) and C_6 corresponds to a $\pi/3$ rotation about the Z-axis

time-reversal symmetry, the system belongs to the unitary or A class and there is an integer invariant that characterizes the ground states of the Hamiltonian. This is the quantum anomalous Hall or Chern insulator proposed by Haldane in 1988 [21], and it is achieved by introducing complex second-neighbour hoppings. Finally, we can take into account that electrons are actually spin-1/2 particles and if time-reversal symmetry is preserved but it is such that $T^2 = -1$, then the system is in the symplectic or AII class and there is a $\mathbb{Z}_2$ index to classify the different ground states. This is the quantum spin Hall or $\mathbb{Z}_2$ topological insulator, proposed by Kane and Mele in 2005 [22, 44] and it relies on spin-orbit coupling. Each of these discoveries would deserve far more than a single section for their review and the reader is referred to the references cited above or those found in Refs. [45–50]. We will discuss however the $\mathbb{Z}_2$ topological insulator in three dimensions after our discussion of graphene.

Although we could use $\boldsymbol{k} \cdot \boldsymbol{p}$ theory to obtain the low-energy description of graphene [19, 37, 51], we will make use of the method of invariants in its simplest form. That is, we will focus on the terms to lowest order in momenta and will not include the effects of strain, electric or magnetic fields. A detailed account of the method applied to the full problem can be found in Ref. [27]. In order to explore the symmetries, we shall consider the graphene lattice to be rotated by $\pi/2$ with respect to that shown in Fig. 2.5b, so that our results coincide with those presented in Ref. [52]. Hence, the Wigner–Seitz unit cell and the Brillouin zone now look as in Fig. 2.8.

Momenta closest to the Fermi energy are the $\boldsymbol{K}$ and $\boldsymbol{K}'$ points, so we shall use as a basis $\left\{\psi_A^K(\boldsymbol{r}), \psi_B^K(\boldsymbol{r}), \psi_B^{K'}(\boldsymbol{r}), \psi_A^{K'}(\boldsymbol{r})\right\}$. In the following, we will not consider intervalley scattering, that is, off-diagonal blocks in the Hamiltonian will be set to zero, for simplicity. Therefore, we can write the Hamiltonian in that basis as follows

$$\mathcal{H}(\boldsymbol{k}) = \begin{pmatrix} H_K(\boldsymbol{k}) & \boldsymbol{0}_2 \\ \boldsymbol{0}_2 & H_{K'}(\boldsymbol{k}) \end{pmatrix}, \tag{2.2.18}$$

where $H_{K(K')}(\boldsymbol{k})$ are 2×2 blocks that we have to determine exploiting the symmetries of graphene and $\boldsymbol{0}_n$ is the n-dimensional null matrix. Taking into account the orientation we have chosen for graphene, reflection about the XZ-plane takes A to B, $(k_x, k_y) \to (k_x, -k_y)$ and it leaves the K and K' points untouched [see Fig. 2.8]. Therefore, in this basis we can choose

$$\mathcal{D}(R_x) = \tau_0 \otimes \sigma_x\ , \tag{2.2.19}$$

where the Pauli matrices τ_i and σ_i act upon the valley and sublattice degrees of freedom, respectively, and the subscript 0 indicates the identity matrix, that is, $\tau_0 = \sigma_0 = \mathbb{1}_2$. Therefore, the Hamiltonian must satisfy

$$\mathcal{D}(R_x)\mathcal{H}(k_x, k_y)\mathcal{D}^{-1}(R_x) = \mathcal{H}(k_x, -k_y)\ . \tag{2.2.20}$$

Another symmetry is rotation about the Z-axis of $\pi/3$, C_6. That symmetry exchanges A and B, K and K' and rotates $\boldsymbol{k}$ as follows

$$(k_x, k_y) \to \left(k_x \cos\varphi + k_y \sin\varphi, -k_x \sin\varphi + k_y \cos\varphi\right)\ , \tag{2.2.21}$$

particularized to $\varphi = \pi/3$. It is easier to express the rotation of $\boldsymbol{k}$ if we express it instead as (k_+, k_-), where $k_\pm = k_x \pm \mathrm{i}\, k_y$. Indeed, after the rotation, $k_\pm$ change to $\exp(\mp \mathrm{i}\,\pi/3)k_\pm$. Since we have to exchange K and K', we can use an operator of the form $\tau_x \otimes \sigma_0$. Notice that this operator also swaps A and B in the basis we have chosen. However, it is not swapping A and B at the same K point, but rather that of the K point with the other of the K' point. In order to keep the symmetry, we must add a phase factor of $\exp(\pm\mathrm{i}\, 2\pi/3)$ accordingly. We do so with the operator $\tau_0 \otimes \exp\left[\mathrm{i}\,(2\pi/3)\sigma_z\right]$. Therefore, the action of C_6 is captured by the following representation [52]

$$\mathcal{D}(C_6) = \tau_x \otimes \exp\left[\mathrm{i}\,(2\pi/3)\sigma_z\right]\ , \tag{2.2.22}$$

and it is such that

$$\mathcal{D}(C_6)\mathcal{H}(k_+, k_-)\mathcal{D}^{-1}(C_6) = \mathcal{H}\left(e^{-\mathrm{i}\pi/3}k_+, e^{\mathrm{i}\pi/3}k_-\right)\ . \tag{2.2.23}$$

Finally, we must consider time-reversal symmetry. Since we are considering spinless particles (or, rather, spin polarized), it must be such that it squares to the identity. One could be tempted to choose again $\Theta = \mathcal{K}$. However, this is not the proper operator in the basis we have chosen [27]. We know that more generally $\Theta = U\mathcal{K}$, where U is a unitary operator, fixed by the chosen representation. On the one hand, $\boldsymbol{K}$ and $\boldsymbol{K}' = -\boldsymbol{K}$ are time-reversed partners, so the operator must exchange the valleys. Just like with our discussion of the previous symmetry, we do this with $\tau_x \otimes \sigma_0$. On the other hand, taking into account the ordering in the basis, after this operation we must also exchange the A and B sublattices with an operator $\tau_0 \otimes \sigma_x$. As a result, the unitary operator U is given by $U = \tau_x \otimes \sigma_x$. Hence, Θ is given by

$$\Theta = (\tau_x \otimes \sigma_x)\mathcal{K} , \tag{2.2.24}$$

which squares to the identity. The action of time-reversal on $\boldsymbol{k}$ is to invert it to $-\boldsymbol{k}$, as we already know, and therefore the action on the Hamiltonian is such that [cf. Eq. (2.1.40)]

$$\Theta\mathcal{H}(\boldsymbol{k})\Theta^{-1} = \mathcal{H}(-\boldsymbol{k}) . \tag{2.2.25}$$

With this in mind, let us explore the restrictions imposed by these symmetries on the matrix elements of the Hamiltonian. Reflection is probably the easiest, since it is diagonal in the valley subspace and leads to the following two equations

$$\sigma_x H_{K(K')}(k_x, k_y)\sigma_x = H_{K(K')}(k_x, -k_y) . \tag{2.2.26}$$

This in turn implies the following restrictions on the matrix elements of $H_{K(K')}$, which we shall denote as $h_{ij}^{K(K')}$,

$$h_{22}^{K(K')}(k_x, k_y) = h_{11}^{K(K')}(k_x, -k_y) \quad \overline{h}_{12}^{K(K')}(k_x, k_y) = h_{12}^{K(K')}(k_x, -k_y) , \tag{2.2.27}$$

where we have used the overline in h to express complex conjugation, that is, $\overline{h} \equiv h^*$. As a result, the K and K' blocks take the form

$$H_{K(K')}(k_x, k_y) = \begin{pmatrix} h_{11}^{K(K')}(k_x, k_y) & h_{12}^{K(K')}(k_x, k_y) \\ h_{12}^{K(K')}(k_x, -k_y) & h_{11}^{K(K')}(k_x, -k_y) \end{pmatrix} . \tag{2.2.28}$$

Time-reversal symmetry requires on the other hand that

$$\sigma_x H_{K'}(k_x, k_y)\sigma_x = H_K^*(-k_x, -k_y) . \tag{2.2.29}$$

For the matrix elements of the blocks this implies that

$$h_{11}^{K'}(k_x, k_y) = h_{11}^{K}(-k_x, k_y) , \qquad h_{12}^{K'}(k_x, k_y) = h_{12}^{K}(-k_x, -k_y) . \tag{2.2.30}$$

Therefore, the two blocks can be written as follows

$$H_K(k_x, k_y) = \begin{pmatrix} h_{11}^{K}(k_x, k_y) & h_{12}^{K}(k_x, k_y) \\ h_{12}^{K}(k_x, -k_y) & h_{11}^{K}(k_x, -k_y) \end{pmatrix} , \tag{2.2.31a}$$

$$H_{K'}(k_x, k_y) = \begin{pmatrix} h_{11}^{K}(-k_x, k_y) & h_{12}^{K}(-k_x, -k_y) \\ h_{12}^{K}(-k_x, k_y) & h_{11}^{K}(-k_x, -k_y) \end{pmatrix} . \tag{2.2.31b}$$

In order to explore the C_6 rotational symmetry, it is convenient to express the two blocks in terms of $k_\pm$. Hence, we will make the correspondence $(k_x, k_y) \to (k_+, k_-)$. The two blocks would then take the form

$$H_K(k_+, k_-) = \begin{pmatrix} h_{11}^K(k_+, k_-) & h_{12}^K(k_+, k_-) \\ h_{12}^K(k_-, k_+) & h_{11}^K(k_-, k_+) \end{pmatrix}, \tag{2.2.32a}$$

$$H_{K'}(k_+, k_-) = \begin{pmatrix} h_{11}^K(-k_-, -k_+) & h_{12}^K(-k_+, -k_-) \\ h_{12}^K(-k_-, -k_+) & h_{11}^K(-k_+, -k_-) \end{pmatrix}. \tag{2.2.32b}$$

If we denote by $\mathcal{R} = \exp\left[\mathrm{i}\,(2\pi/3)\sigma_z\right]$, then the C_6 rotational symmetry implies that

$$\mathcal{R} H_{K'}(k_+, k_-)\mathcal{R}^{-1} = H_K\left(e^{-\mathrm{i}\pi/3}k_+, e^{\mathrm{i}\pi/3}k_-\right). \tag{2.2.33}$$

This requires for the matrix elements to satisfy

$$h_{11}^K(k_+, k_-) = h_{11}^K\left(-e^{\mathrm{i}\pi/3}k_-, -e^{-\mathrm{i}\pi/3}k_+\right), \tag{2.2.34a}$$

$$h_{12}(k_+, k_-) = -e^{-\mathrm{i}\pi/3}h_{12}(-e^{-\mathrm{i}\pi/3}k_+, -e^{\mathrm{i}\pi/3}k_-). \tag{2.2.34b}$$

With these two requirements, we can obtain the form of these matrix elements to lowest order in k_x, k_y. For the diagonal elements, it is clear that the only allowed terms to lowest order are

$$h_{11}^K(k_+, k_-) \simeq a_{11} + b_{11}k_+k_- = a_{11} + b_{11}k^2 \equiv \varepsilon(\boldsymbol{k}), \tag{2.2.35}$$

where $k = |\boldsymbol{k}|$. Since those terms are in the diagonal of the Hamiltonian, the coefficients have to be real to ensure Hermiticity. On the other hand, for the off-diagonal element h_{12}, we can observe that the only allowed terms that preserve the symmetries are proportional to k_- to first order

$$h_{12}^K(k_+, k_-) \simeq a_{12}k_-. \tag{2.2.36}$$

From the condition (2.2.27), we obtain that a_{12} has to be real. In summary, the two blocks would now look like so

$$H_K(k_+, k_-) = \begin{pmatrix} \varepsilon(\boldsymbol{k}) & a_{12}k_- \\ a_{12}k_+ & \varepsilon(\boldsymbol{k}) \end{pmatrix} = \varepsilon(\boldsymbol{k})\sigma_0 + a_{12}\left(\sigma_x k_x + \sigma_y k_y\right), \tag{2.2.37a}$$

$$H_{K'}(k_+, k_-) = \begin{pmatrix} \varepsilon(\boldsymbol{k}) & -a_{12}k_- \\ -a_{12}k_+ & \varepsilon(\boldsymbol{k}) \end{pmatrix} = \varepsilon(\boldsymbol{k})\sigma_0 - a_{12}\left(\sigma_x k_x + \sigma_y k_y\right). \tag{2.2.37b}$$

The full Hamiltonian can then be written in a very succinct way

$$\mathcal{H}(\boldsymbol{k}) = \varepsilon(\boldsymbol{k})\tau_0 \otimes \sigma_0 + \tau_z \otimes (a_{12}\boldsymbol{\sigma}_\perp \cdot \boldsymbol{k}), \tag{2.2.38}$$

where $\boldsymbol{\sigma}_\perp = (\sigma_x, \sigma_y)$. To lowest order, $\varepsilon(\boldsymbol{k})$ is independent of $\boldsymbol{k}$ and can be absorbed in the definition of the energy. In such a case, we have that each valley is described by a Weyl equation (i.e. a massless Dirac equation), and therefore the spectrum is that of a Dirac cone, $E(k) = \pm a_{12}k$. Thus, a_{12} is to be interpreted in the present context as the Fermi velocity. Finally, the Hamiltonian is written as follows

$$\mathcal{H}(\boldsymbol{k}) = v_F \tau_z \otimes (\boldsymbol{\sigma}_\perp \cdot \boldsymbol{k}) \,. \tag{2.2.39}$$

It must be noted that upon application of external fields or strain, a lot more terms must be added to the Hamiltonian [27], many of which depend on spin, which we have neglected. In fact, the quantum spin Hall effect includes one such term which has driven the revolution in the field of topological insulators [22, 44]. In any case, however, we shall keep our system to be spinless, so that we will not consider such terms.

2.2.2 *Topological Insulators Low-Energy Hamiltonian*

The last model that we will explore using the method of invariants is that of three-dimensional topological insulators. In particular, we shall consider those presented in Ref. [53], namely Bi_2Se_3, Bi_2Te_3 and Sb_2Te_3. These materials are insulators and they present time-reversal symmetry (i.e. they are non-magnetic). Considering spin, the time-reversal operator then satisfies the restriction for spin-1/2 systems, that is, $\Theta^2 = -1$. We shall begin by exploring the main consequence of having $\Theta^2 = -1$ in a Bloch Hamiltonian. But first, let us observe what is the prediction according to the Altland–Zirnbauer classification (see Table 2.1) [29–33]. Considering only time-reversal symmetry and the fact that it squares to minus the identity, we can only place such gapped phases in the AII class. In two or three dimensions, the classifications predicts that gapped ground states can be classified according to a nontrivial $\mathbb{Z}_2$ index. We shall see that this is indeed the case and will provide a method to compute such invariants. Remember that a Bloch Hamiltonian satisfies

$$\Theta\mathcal{H}(\boldsymbol{k})\Theta^{-1} = \mathcal{H}(-\boldsymbol{k}) \,. \tag{2.2.40}$$

Taking into account that the Brillouin zone is a torus, there are special points where $\boldsymbol{k}$ and $-\boldsymbol{k}$ are equivalent, that is, they are related by a reciprocal lattice vector. Those *time-reversal invariant momenta* (TRIM), which we shall denote as q_i, where i runs through all possible TRIM, are then such that

$$\Theta\mathcal{H}(q_i)\Theta^{-1} = \mathcal{H}(q_i) \,. \tag{2.2.41}$$

A particular example is the Γ point, $q_i = 0$. In two dimensions, there are four TRIM, whereas in three dimensions there are eight TRIM [49, 54]. As we can observe, at those particular points, the Hamiltonian commutes with Θ, meaning that $|\psi\rangle$ and

$\Theta|\psi\rangle$ are eigenstates of the Hamiltonian with the same energy. The question now is if these states are actually the same state or if they are degenerate. If they are the same state, then we can write $\Theta|\psi\rangle = c|\psi\rangle$, where c is a complex number. However, if we apply Θ and take into account that it is antilinear, we find $\Theta^2|\psi\rangle = c^*\Theta|\psi\rangle = |c|^2|\psi\rangle$. Now, for spin-1/2 systems, $\Theta^2 = -1$, implying that $|c|^2 = -1$, which is obviously impossible. Therefore, $|\psi\rangle$ and $\Theta|\psi\rangle$ are actually degenerate at the TRIM. This is known as *Kramers' degeneracy* and the two states are commonly referred to as *Kramers' pairs*. Away from the TRIM, the degeneracy splits. With this in mind, Kane and Mele derived a $\mathbb{Z}_2$ invariant in terms of the Pfaffian of an antisymmetric matrix by taking scalar products between states $|\psi\rangle$ and $\Theta|\psi\rangle$ at the TRIM [44]. We shall not be interested in such a computation and will limit ourselves to a rather simpler version of the invariant which applies to systems with inversion symmetry, such as the ones mentioned before. In fact, we will not detail the derivation of such calculations, but rather give the final result, the details of which can be found in the original paper by Fu and Kane [54]. Since we are considering insulators (gapped phases), there are a number of fully occupied bands, let us call it $2N$, where the extra factor of 2 comes from Kramers' degeneracy. It must be restated that there is no Kramers' degeneracy out of the TRIM, but it is convenient to say that we have $2N$ occupied bands. It is also common to find in the literature the name *Kramers' pairs of bands* to denote those bands that arise from the degenerate Kramers' pair at the TRIM. The interesting bit of considering inversion symmetry is that it allows us to label the eigenstates by their parity. Let $\xi_{2m}(q_i)$ be the parity of the $2m$th band at the TRIM q_i, which coincides with the parity of its Kramers' partner [49], $\xi_{2m-1}(q_i)$. Fu and Kane proposed that the $\mathbb{Z}_2$ invariant, ν, can be obtained from the product of these such parities for the $2N$ occupied bands as follows

$$(-1)^\nu = \prod_i \delta_i\,, \qquad \delta_i = \prod_{m=1}^{N} \xi_{2m}(q_i)\,. \tag{2.2.42}$$

This expression has a direct implication: if a system undergoes a band inversion where the parities of the bands are inverted at an odd number of TRIM with respect to a system with opposite ordering, then $\nu \to \nu + 1$ in going from one system to the other, meaning that they belong to different topological sectors. If inversion occurs at an even number of TRIM, then ν does not change at all and the two systems belong to the same topological sector.

With this knowledge, let us explore Bi_2Se_3, Bi_2Te_3, Sb_2Se_3 and Sb_2Te_3. In order to follow Refs. [53, 55], we shall consider Bi_2Se_3 as a representative of such a family of materials, all of which present a similar crystal structure, shown in Fig. 2.9. The unit cell contains five atoms, two of which are equivalent atoms of Se (denoted by Se1 and Se1' and both with the same colouring), two equivalent atoms of Bi (Bi1 and Bi1') and one nonequivalent atom of Se (dubbed Se2). Those atoms arrange themselves into planes forming triangular lattices. These planes are strongly bonded forming quintuple layers. Coupling between quintuple layers is weaker, since it is a Van der Waals bond. There are various symmetries that we will explore later on, but for

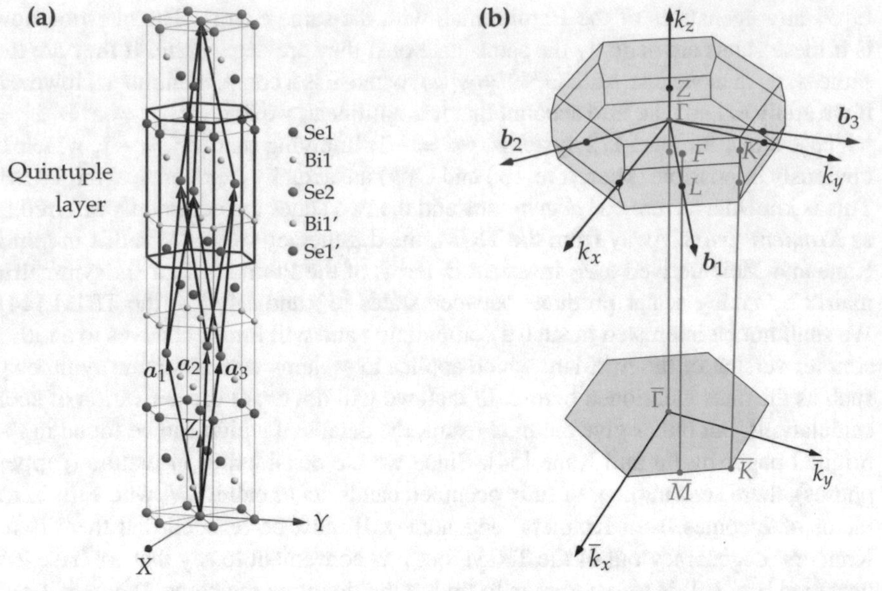

Fig. 2.9 Crystallography of Bi_2Se_3. **a** The crystal lattice is orthorhombic, the primitive cell is the parallelepiped enclosed by the arrows, which denote the primitive vectors $\boldsymbol{a}_1, \boldsymbol{a}_2, \boldsymbol{a}_3$. Each primitive cell contains 5 atoms, two equivalent Se atoms (Se1 and Se1'), one inequivalent Se atom (Se2) and two equivalent Bi atoms (Bi and Bi'). Each type of atom is arranged forming layers that consist on triangular lattices parallel to the XY plane. Bonding between different layers is such that it is possible to form quintuple layers of strongly bonded layers, each quintuple layer bonded to the next via weak Van der Waals bonds. **b** First Brillouin zone of the bulk crystal (top) and the (111) surface Brillouin zone (bottom), displaying the reciprocal lattice vectors $\boldsymbol{b}_1, \boldsymbol{b}_2, \boldsymbol{b}_3$ and the high-symmetry points [55]

Table 2.2 $\mathbb{Z}_2$ invariant in antimony and bismuth chalcogenides. Parities of fourteen bands at the Γ-point below the Fermi level are shown on the middle column. The $\mathbb{Z}_2$ invariant is simply obtained from the product of parities at the Γ-point and Eq. (2.2.42), since it is the only TRIM where parity inversion occurs [53]

Material	Parities	ν
Sb_2Se_3	+ − + − + − + − + − + − − −	+1
Sb_2Te_3	+ − + − + + − + − + − − − +	−1
Bi_2Se_3	+ − + − + − + − + − + − − +	−1
Bi_2Te_3	+ − + − + − + + − + − − − +	−1

now we can observe that taking Se2 as the inversion center [55], the system displays inversion symmetry and we can use the results that we discussed earlier regarding the $\mathbb{Z}_2$ invariant. Ab-initio density functional theory calculations performed in Ref. [53] have shown that the product of parities for the fourteen occupied bands at the Γ point is −1 for Bi_2Se_3, Bi_2Te_3 and Sb_2Te_3, whereas it is +1 at Sb_2Se_3 [cf. Table 2.2].

The ab-initio calculations of Ref. [53] show that in Bi_2Se_3, Bi_2Te_3 and Sb_2Te_3, before turning on the spin-orbit interaction, the product of parities renders $+1$, as in Sb_2Se_3. As a result, the spin-orbit interaction is responsible for a band inversion at the Γ-point. In fact, the authors of Ref. [53] have been able to observe that inversion occurs exclusively at the Γ-point [55], which implies from Table 2.2 and Eq. (2.2.42) that the $\mathbb{Z}_2$ invariant is $\nu = 1$ for Bi_2Se_3, Bi_2Te_3 and Sb_2Te_3 and $\nu = 0$ for Sb_2Se_3. Hence, the first three chalcogenides are topological insulators and the last one is a trivial insulator. Moreover, the ab-initio calculations of Ref. [53] also show that, as expected, the systems with non-trivial topology display topological surface states, whereas Sb_2Se_3 does not.

In order to obtain the low-energy Hamiltonian around the Γ point using the method of invariants, we must explore the symmetries in a given basis. Finding such a basis, however, is not as straightforward as in graphene, and it pays off to consider the evolution of the orbitals from Bi and Se as different terms in the Hamiltonian are turned on, schematically. This is shown in Fig. 2.10 for the levels at the Γ-point. Although Fig. 2.10 is schematic, it can be shown quantitatively that the evolution of atomic levels occurs as displayed in such a figure, as discussed in [55]. However, we shall consider only the more intuitive picture to understand the appearance of topology in these systems. It is important to recall that bonding within a quintuple layer is strong, whereas there is weak Van der Waals between quintuple layers. Therefore, one can focus on a single quintuple layer to understand the evolution of the isolated atomic energy levels as different effects are included in the system. These effects are included in progression with respect to their associated energy scale, a quantitative discussion of which is found in [55]. The electron configuration of Bi is $6s^2 6p^3$ and that of Se is $4s^2 4p^4$, which implies that one can focus on the outermost p orbitals and disregard the s orbitals. Each Se and Bi atom contributes three p orbitals, p_x, p_y, p_z. Since there is a total of five atoms within a unit cell, there are 15 orbitals in total, 9 from Se and 6 from Bi. As it can be observed from Fig. 2.9, layers of Bi and Se alternate within a quintuple layer, meaning that the strongest coupling occurs due to chemical bonding between neighbouring layers of Bi and Se. This coupling leads to hybridization and, as a result, to level repulsion of the atomic orbitals and corresponds to the first stage in Fig. 2.10. The resulting six hybrid orbitals of Bi are denoted as $B_{x,y,z}$ and $B'_{x,y,z}$, where x, y, z corresponds to p_x, p_y, p_z, and those of Se are denoted as $S_{x,y,z}$, $S'_{x,y,z}$ and $S0_{x,y,z}$. In order to take advantage of inversion symmetry, it is convenient to rearrange the p orbitals of the three Se atoms so as to obtain states of well-defined parity [55], thereby leading to two odd states, $P0^-_{x,y,z}$ and $P2^-_{x,y,z}$, and an even state, $P2^+_{x,y,z}$ for each p orbital, where $\pm$ indicates the parity eigenvalue. The same applies to Bi, leading to an even, $P1^+_{x,y,z}$ and an odd state $P1^-_{x,y,z}$ for each p orbital. It is important to bear in mind that states of opposite parity cannot be coupled as long as inversion symmetry is preserved and parity is a well-defined quantity. We will therefore observe that there are no $\pm$ signs simultaneously on a given level. States of opposite parity split, making those of even parity move downwards in energy and those of odd parity move upwards, similar to the formation of bonding and antibonding states. This corresponds to the second stage in Fig. 2.10. States closest to the Fermi level are therefore $P1^+_{x,y,z}$ and $P2^-_{x,y,z}$, as shown in the figure. We focus on these and disregard all other states in what follows. The next stage corresponds to the fact that the system is not spherically symmetric, that is, the Z direction is different from the X and Y directions, resulting in *crystal field splitting*. That is, the degeneracy between x, y and z orbitals splits apart due to the interaction with the field created by neighbouring atoms, as shown in the third stage of Fig. 2.10. After splitting, the two states closest to the Fermi level are two p_z levels of opposite parity, $P1_z^+$ and $P2_z^-$. Finally, the last stage corresponds to turning on the spin-orbit coupling. Up to now, we have not taken into account the fact that all these orbitals are doubly degenerate due to spin since we had not included any terms that couple spin to other degrees of freedom. However, spin-orbit coupling causes the orbital angular momentum and spin to couple, so that the four orbitals in $P1_{x,y}^+$ and $P2_{x,y}^-$ evolve into states with well defined total angular momentum, $P1_{x\pm\mathrm{i}\,y,\uparrow\downarrow}^+$ and $P2_{x\pm\mathrm{i}\,y,\uparrow\downarrow}^+$, where $x \pm \mathrm{i}\,y$ denotes $p_x \pm \mathrm{i}\,p_y$ and $\uparrow\downarrow$ denotes the spin eigenvalue. If m_J denotes the eigenvalue of the Z-component of total angular momentum, states of the same m_J will couple due to spin-orbit coupling, leading to level repulsion. This implies that $P1_{x+\mathrm{i}\,y,\downarrow}^+$ and $P1_{z,\uparrow}^+$, which have $m_J = 1/2$, will couple into two new levels which will repel. Each level will have a stronger contribution from either $P1_{x+\mathrm{i}\,y,\downarrow}^+$ or $P1_{z,\uparrow}^+$. In order to ease notation, we shall denote the level with more $P1_{z,\uparrow}^+$ weight as $P1_{z,1/2}^+$ and that with more $P1_{x+\mathrm{i}\,y,\downarrow}^+$ weight as $P1_{1/2}^+$. The same applies to $P1_{x-\mathrm{i}\,y,\uparrow}^+$ and $P1_{z,\downarrow}^+$ with $m_J = -1/2$, which will couple into $P1_{z,-1/2}^+$ and $P1_{-1/2}^+$. Similarly, $P2_{x+\mathrm{i}\,y,\downarrow\uparrow}^-$ and $P2_{z,\uparrow\downarrow}^-$, that will couple into $P2_{\pm 1/2}^-$ and $P2_{z,\pm 1/2}^-$. It is important to remember that states from Bi with the same m_J as those of Se will not couple since they have opposite parity. The states $P1_{x\pm\mathrm{i}\,y,\uparrow\downarrow}^+$ and $P2_{x\pm\mathrm{i}\,y,\uparrow\downarrow}^-$ with $m_J = \pm 3/2$ do not couple with other states, so we will just relabel them as $P1_{\pm,\pm 3/2}^+$ and $P2_{\pm,\pm 3/2}^-$ respectively. On the other hand, spin-orbit coupling also breaks the degeneracy between $J = 3/2$ and $J = 1/2$, leading to a splitting between $P2_{\pm,\pm 3/2}^+$ and $P2_{\pm,\pm 1/2}^+$ pushing the states with $J = 3/2$ upwards in energy and lowering the states with $J = 1/2$. Identically happens for $P1_{\pm,\pm 3/2}^-$ and $P1_{\pm,\pm 1/2}^-$. Notice, however, that a double degeneracy is still present since states of opposite m_J do not split since angular momentum in the Z-direction is still a good quantum number [55]. The net result of this stage is that, due to spin-orbit coupling, levels of opposite parity closest to the Fermi energy are inverted and, in consequence, the total parity below the Fermi level changes sign. Since we are performing this analysis at the Γ point (a TRIM in the Brillouin zone) and this inversion occurs exclusively at this point, the system becomes topologically non-trivial, as discussed above. Also notice that, as it should be, all states are still doubly degenerate due to Kramers' theorem. This same evolution of the energy levels occurs for Bi_2Se_3, Bi_2Te_3 and Sb_2Te_3, but in Sb_2Se_3 spin-orbit coupling is not strong enough (Se is a much lighter element compared to Te) and there is no parity inversion, so the system remains a trivial insulator.

Knowing that states closest to the Fermi level are $P1_{z,\pm 1/2}^+$ and $P2_{z,\pm 1/2}^-$ we can construct our low-energy Hamiltonian using a basis containing those four orbitals. It is common to abuse of terminology and refer to spin instead of total angular momentum [20, 53, 56]. This is understandable, since these two states come mainly

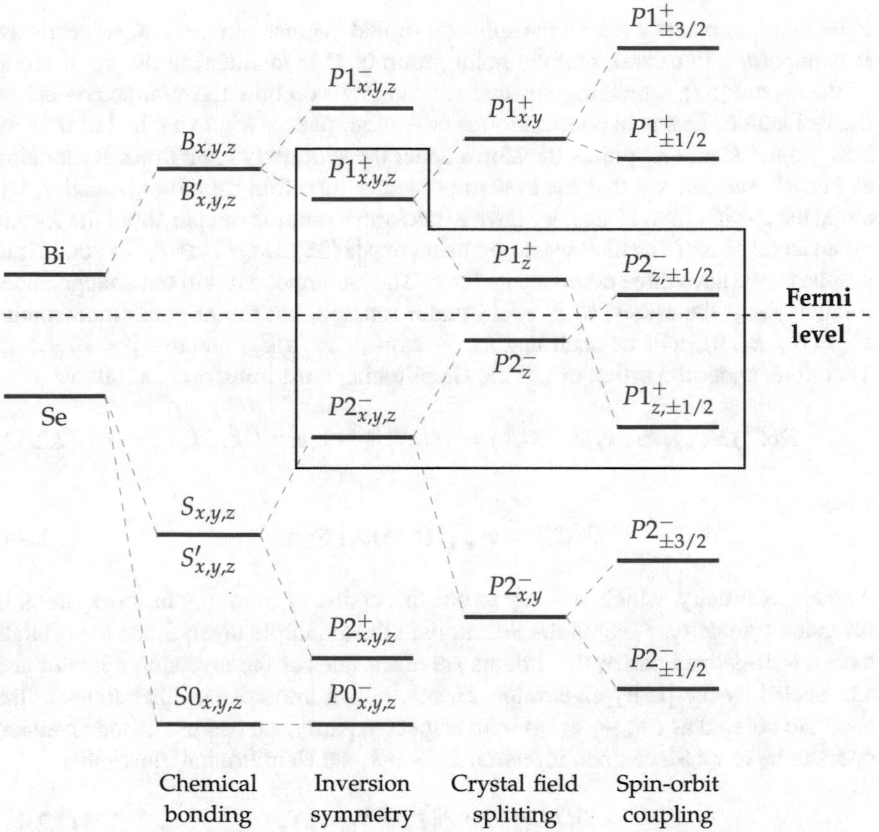

Fig. 2.10 Evolution of atomic levels into band edges at the Γ point in Bi_2Se_3. Chemical bonding makes p orbitals from Bi and Se to hybridize; inversion symmetry allows us to label states according to their parity and states of opposite parity split; crystal field splitting occurs due to anisotropy between the X, Y and Z directions; spin-orbit coupling couples spin and angular momentum, mixing states of same total angular momentum and leading to level repulsion. A black frame encloses those levels closest to the Fermi level (dashed blue line). The net result is that two there is an inversion of parity on the last stage [55]

from $P1^{+}_{z,\uparrow(\downarrow)}$ and $P2^{-}_{z,\uparrow(\downarrow)}$. Indeed, the crystal field splitting is quite strong, thereby separating very much in energy the states $P1^{+}_{z,\uparrow(\downarrow)}$ and $P2^{-}_{z,\uparrow(\downarrow)}$ from those they couple to due to spin-orbit coupling, as shown in the third stage of Fig. 2.10. Hence, after spin orbit coupling, even though there will be an increased repulsion, $P1^{+}_{z,\pm 1/2}$ and $P2^{-}_{z,\pm 1/2}$ will come mostly from $P1^{+}_{z,\uparrow(\downarrow)}$ and $P2^{-}_{z,\uparrow(\downarrow)}$ respectively. This is why they are linked by dashed lines in Fig. 2.10 in going from the third to the last stage. We will therefore abuse of terminology as well and denote them hereafter by $P1^{+}_{z,\uparrow(\downarrow)}$ and $P2^{-}_{z,\uparrow(\downarrow)}$ respectively. We shall then use as a basis $\{P1^{+}_{z,\uparrow}, P2^{-}_{z,\uparrow}, P1^{+}_{z,\downarrow}, P2^{-}_{z,\downarrow}\}$. Equivalently, we will say that we are working in the spin-orbital basis, $\{\sigma, \tau\}$, so that

Pauli matrices σ and τ act on the spin and orbital degrees of freedom, respectively. It is important to remark that the point group of Γ is identical to the point group of the crystal [57], which implies that we can focus on how the symmetries act on the real lattice. This is in contrast to the case of graphene, where we had to observe how did the K and K' points transform under the symmetry operations. By looking at Fig. 2.9, we can see that the system possesses threefold rotation symmetry, C_3^z, along the Z-direction. Hence, we have to perform a rotation of spin about the Z-axis by an angle of $2\pi/3$, which we do by using $\exp\left[\mathrm{i}\,(2\pi/3)\sigma_z/2\right] \otimes \tau_0$. In contrast to graphene, we have three components for $\boldsymbol{k}$. The z-component will not change under rotation about the z-axis, so $k_z \to k_z$ under rotation, but the in-plane momentum, $\boldsymbol{k}_\perp = (k_x, k_y, 0)$, will be such that $k_\pm \to \exp(\mp \mathrm{i}\, 2\pi/3) k_\pm$, where $k_\pm = k_x \pm \mathrm{i}\, k_y$. Therefore, under the action of C_3^z, the Hamiltonian must transform as follows

$$\mathcal{D}(C_3^z)\mathcal{H}(k_+, k_-, k_z)\mathcal{D}^{-1}(C_3^z) = \mathcal{H}(e^{-\mathrm{i}\,2\pi/3}k_+, e^{\mathrm{i}\,2\pi/3}k_-, k_z) , \tag{2.2.43}$$

where

$$\mathcal{D}(C_3^z) = \exp\left[\mathrm{i}\,(\pi/3)\sigma_z\right] \otimes \tau_0 . \tag{2.2.44}$$

Another symmetry, which was key to our discussion of topology in this system, is inversion symmetry, I, about the Se2 atom in the quintuple layer. Since the orbitals have a well-defined parity, the orbitals are eigenstates of the inversion operator and are labeled by the parity eigenvalue. Hence, taking into account that states in the basis are ordered as $(+, -, +, -)$ with respect to parity, we can use as the inversion operator $\sigma_0 \otimes \tau_z$. Since upon inversion $\boldsymbol{k} \to -\boldsymbol{k}$, the Hamiltonian must satisfy

$$\mathcal{D}(I)\mathcal{H}(\boldsymbol{k})\mathcal{D}^{-1}(I) = \mathcal{H}(-\boldsymbol{k}) , \tag{2.2.45}$$

with

$$\mathcal{D}(I) = \sigma_0 \otimes \tau_z . \tag{2.2.46}$$

The crystal also exhibits twofold rotational symmetry about the Y axis. We therefore have to rotate spin around the Y-axis an angle of π, which is done via $\exp\left(\mathrm{i}\,\pi\sigma_y/2\right) \otimes \tau_0 = \mathrm{i}\,\sigma_y \otimes \tau_0$. This symmetry takes $(k_x, k_y, k_z) \to (-k_x, k_y, -k_z)$. Therefore, the Hamiltonian must satisfy

$$\mathcal{D}(C_2^y)\mathcal{H}(k_x, k_y, k_z)\mathcal{D}^{-1}(C_2^y) = \mathcal{H}(-k_x, k_y, -k_z) , \tag{2.2.47}$$

where

$$\mathcal{D}(C_2^y) = \mathrm{i}\,\sigma_y \otimes \tau_0 . \tag{2.2.48}$$

Finally, there is time-reversal symmetry. In contrast to the case of graphene where one has to find a unitary operator that exchanges the two valleys, the unitary operator in the orbital subspace can in this case be chosen as the identity, like in the SSH model. In the spin subspace we need to include the $\mathrm{i}\,\sigma_y$ term. Therefore, the unitary operator in time-reversal symmetry would be $\mathrm{i}\,\sigma_y \otimes \tau_0$. Under time-reversal, $\boldsymbol{k} \to -\boldsymbol{k}$, so the

Hamiltonian must satisfy

$$\Theta\mathcal{H}(\boldsymbol{k})\Theta^{-1} = \mathcal{H}(-\boldsymbol{k})\,, \tag{2.2.49}$$

where

$$\Theta = \left(\mathrm{i}\,\sigma_y \otimes \tau_0\right)\mathcal{K}\,. \tag{2.2.50}$$

With all this knowledge, we can proceed to obtain our low-energy Hamiltonian, in a similar manner as to what we did in graphene. Let us denote the four blocks of the Hamiltonian as $h_{ij}(\boldsymbol{k})$ with $i, j = 1, 2$, $h_{ii} = h_{ii}^\dagger$ and $h_{21} = h_{12}^\dagger$. On the one hand, time-reversal symmetry requires that

$$h_{22}(\boldsymbol{k}) = h_{11}^*(-\boldsymbol{k})\,, \qquad h_{12}(\boldsymbol{k}) = -h_{12}^T(-\boldsymbol{k})\,. \tag{2.2.51}$$

Hence, we must only worry about two of the three independent blocks, a property that we will take into account when studying the other symmetries. C_3^z rotation symmetry requires that

$$\begin{aligned} h_{11}(k_+, k_-, k_z) &= h_{11}(e^{-\mathrm{i}2\pi/3}k_+, e^{\mathrm{i}2\pi/3}k_-, k_z)\,,\\ h_{12}(k_+, k_-, k_z) &= e^{-\mathrm{i}2\pi/3}h_{11}(e^{-\mathrm{i}2\pi/3}k_+, e^{\mathrm{i}2\pi/3}k_-, k_z)\,. \end{aligned} \tag{2.2.52}$$

On the other hand, inversion symmetry implies that

$$\tau_z h_{ij}(\boldsymbol{k})\tau_z = h_{ij}(-\boldsymbol{k})\,. \tag{2.2.53}$$

Finally, C_2^y rotational symmetry imposes the following conditions

$$\begin{aligned} h_{11}(k_x, k_y, k_z) &= h_{11}^*(k_x, -k_y, k_z)\,,\\ h_{12}(k_x, k_y, k_z) &= h_{12}^*(k_x, -k_y, k_z)\,. \end{aligned} \tag{2.2.54}$$

The conditions imposed by C_3^z imply that h_{11} can only have terms to lowest order proportional to k_z, k_+k_-, k_z^2 and h_{12} can have terms proportional to k_-. However, inversion symmetry implies that the diagonal terms of both blocks have to be even in $\boldsymbol{k}$ and the off-diagonal terms have to be odd in $\boldsymbol{k}$. Therefore, h_{11} can have terms proportional to k_+k_- and k_z^2 in the diagonal and proportional to k_z in the off-diagonal, whereas h_{12} can only have terms in the off-diagonal proportional to k_- and it is traceless. Thus,

$$h_{11}(\boldsymbol{k}) = \begin{pmatrix} a_{11} + b_{11}k_+k_- + c_{11}k_z^2 & a_{12}k_z \\ a_{12}^*k_z & a_{22} + b_{22}k_+k_- + c_{22}k_z^2 \end{pmatrix}, \tag{2.2.55}$$

where the coefficients in the diagonal are real. For $h_{12}(\boldsymbol{k})$ we have

$$h_{12}(\boldsymbol{k}) = \begin{pmatrix} 0 & d_{12}k_- \\ e_{12}k_- & 0 \end{pmatrix}. \tag{2.2.56}$$

From C_2^y, we obtain that $a_{12} = a_{12}^*$, $d_{12} = d_{12}^*$ and $e_{12} = e_{12}^*$, so all those coefficients are also real. Finally, time-reversal symmetry requires that $d_{12} = e_{12}$. Hence, we can write the two blocks as follows

$$h_{11}(\boldsymbol{k}) = \begin{pmatrix} a_{11} + b_{11}k_+k_- + c_{11}k_z^2 & a_{12}k_z \\ a_{12}k_z & a_{22} + b_{22}k_+k_- + c_{22}k_z^2 \end{pmatrix}, \tag{2.2.57}$$

$$h_{12}(\boldsymbol{k}) = d_{12}k_-\tau_x , \tag{2.2.58}$$

where all the coefficients are real. Another way of writing $h_{11}(\boldsymbol{k})$ is the following

$$h_{11}(\boldsymbol{k}) = \varepsilon(\boldsymbol{k}) + M(\boldsymbol{k})\tau_z + a_{12}k_z\tau_x , \tag{2.2.59}$$

where

$$\begin{aligned} \varepsilon(\boldsymbol{k}) &= \frac{1}{2}\left[a_{11} + a_{22} + k_\perp^2(b_{11} + b_{22}) + k_z^2(c_{11} + c_{22})\right] , \\ M(\boldsymbol{k}) &= \frac{1}{2}\left[a_{11} - a_{22} + k_\perp^2(b_{11} - b_{22}) + k_z^2(c_{11} - c_{22})\right] . \end{aligned} \tag{2.2.60}$$

In conclusion, the low energy Hamiltonian of this system can be written as follows

$$\mathcal{H}(\boldsymbol{k}) = \begin{pmatrix} \varepsilon(\boldsymbol{k}) + M(\boldsymbol{k}) & a_{12}k_z & 0 & d_{12}k_- \\ a_{12}k_z & \varepsilon(\boldsymbol{k}) - M(\boldsymbol{k}) & d_{12}k_- & 0 \\ 0 & d_{12}k_+ & \varepsilon(\boldsymbol{k}) + M(\boldsymbol{k}) & -a_{12}k_z \\ d_{12}k_+ & 0 & -a_{12}k_z & \varepsilon(\boldsymbol{k}) - M(\boldsymbol{k}) \end{pmatrix} . \tag{2.2.61}$$

In a more compact form, we can also write

$$\mathcal{H}(\boldsymbol{k}) = \varepsilon(\boldsymbol{k})\mathbb{1}_4 + M(\boldsymbol{k})\sigma_0 \otimes \tau_z + a_{12}k_z\sigma_z \otimes \tau_x + d_{12}k_x\sigma_x \otimes \tau_x + d_{12}k_y\sigma_y \otimes \tau_x . \tag{2.2.62}$$

As it is, this equation may not be familiar yet. However, let us rearrange the basis, so that we work in the orbital-spin basis, $\{\tau, \sigma\}$, instead of the spin-orbital basis in which $\mathcal{H}(\boldsymbol{k})$ is currently written. This amounts to exchanging the positions of the operators in the products above, that is,

$$\mathcal{H}(\boldsymbol{k}) = \varepsilon(\boldsymbol{k})\mathbb{1}_4 + M(\boldsymbol{k})\tau_z \otimes \sigma_0 + a_{12}k_z\tau_x \otimes \sigma_z + d_{12}k_x\tau_x \otimes \sigma_x + d_{12}k_y\tau_x \otimes \sigma_y . \tag{2.2.63}$$

This Hamiltonian is nothing but a Dirac Hamiltonian in 3+1 dimensions, with a momentum dependent mass term and an additional term proportional to the identity, $\varepsilon(\boldsymbol{k})$. Recalling that the Dirac matrices are defined as

$$\alpha_i = \tau_x \otimes \sigma_i , \qquad \beta = \tau_z \otimes \sigma_0 , \tag{2.2.64}$$

we can immediately write the low-energy Hamiltonian as follows

$$\mathcal{H}(\boldsymbol{k}) = \varepsilon(\boldsymbol{k})\mathbb{1}_4 + M(\boldsymbol{k})\beta + d_{12}\boldsymbol{\alpha}_\perp \cdot \boldsymbol{k}_\perp + a_{12}\alpha_z k_z \,, \tag{2.2.65}$$

where the subscript $\perp$ in a vector indicates the nullification of the z component, $\boldsymbol{u}_\perp = (u_x, u_y, 0)$. We can then try to give an interpretation for the terms appearing in the Hamiltonian. If we consider $\boldsymbol{k} = 0$, then the Hamiltonian reads

$$\mathcal{H}(0) = \varepsilon(0)\mathbb{1}_4 + M(0)\beta \,. \tag{2.2.66}$$

This Hamiltonian is diagonal and its energies must then correspond directly to the band-edge energies at the Γ point. Hence, $\varepsilon(0) + M(0)$ is the conduction band-edge energy and $\varepsilon(0) - M(0)$ is the valence band-edge energy. Therefore, $2M(0) = E_{\mathrm{G}}$ is the energy gap and $\varepsilon(0) \equiv V_{\mathrm{C}}$ is the position of the gap center. On the other hand, for low momenta, $\varepsilon(\boldsymbol{k})$ and $M(\boldsymbol{k})$ are independent of $\boldsymbol{k}$ and the Hamiltonian strictly becomes a Dirac Hamiltonian with anisotropy in the velocity. That is, the coefficients d_{12} and a_{12} can be interpreted as the equivalent to the speed of light in the Dirac equation. We shall write them as $d_{12} = v_\perp$ and $a_{12} = v_z$. Therefore, to lowest order in $\boldsymbol{k}$ the low-energy Hamiltonian reads

$$\mathcal{H}(\boldsymbol{k}) = v_\perp \boldsymbol{\alpha}_\perp \cdot \boldsymbol{k}_\perp + v_z \alpha_z k_z + \frac{1}{2} E_{\mathrm{G}}\beta + V_{\mathrm{C}}\mathbb{1}_4 \,. \tag{2.2.67}$$

This Hamiltonian is particularly interesting to understand the topology that we discussed earlier. Indeed, if we diagonalize this Hamiltonian, the dispersion is that of massive Dirac fermions

$$E(\boldsymbol{k}) = \pm\sqrt{(v_\perp k_\perp)^2 + (v_z k_z)^2 + \frac{1}{4}E_{\mathrm{G}}^2} \,. \tag{2.2.68}$$

This equation predicts the same gapped ground state for either positive or negative E_{G}. However, since E_{G} is the energy difference between the $P1^+_{z,\uparrow(\downarrow)}$ and $P2^+_{z,\uparrow(\downarrow)}$, having a positive or a negative value of E_{G} leads to topologically distinct behaviour. Indeed, under band inversion it changes sign. Hence, the two ground states cannot be topologically equivalent, even though they have the same energy. This is one of the key features of topological matter, which appeared also in the SSH model: two ground states may have the same energy, but they may be topologically distinct. Therefore, the spectrum alone cannot convey all the physics contained in a given material. Since E_{G} changes sign under band inversion, we may just as well identify the $\mathbb{Z}_2$ index as the sign of E_{G}, that is [20],

$$\mathbb{Z}_2 = \operatorname{sgn}(E_{\mathrm{G}}) \,. \tag{2.2.69}$$

In a similar manner as to the SSH model, we would expect at an interface between two materials of opposite index (say a topological insulator and a vacuum or trivial insulator) the existence of gapless surface states as a result of the bulk-boundary correspondence. We shall see in the next section that this is indeed the case. In fact, those gapless states will be the subject of most of this Thesis.

Finally, before we move on to the next section, it is worth it to say a few more words about this low-energy Hamiltonian. Although it has been recently discovered to describe the aforementioned three-dimensional topological insulators, it was already known to describe a particular kind of so-called *narrow gap semiconductors*: $Pb_{1-x}Sn_xTe$. In this family of ternary compounds, upon variation of the Sn fraction, x, the bands undergo band-inversion at the L-points of the Brillouin zone, leading to an inversion of the L_6^+ and L_6^- band edges, and therefore to an inversion of parity. A low-energy Hamiltonian with exactly the same form as the one we have obtained for the three-dimensional topological insulators was found by Dimmock et al. in 1964 [58] and got further attention for more than 20 years after that [59–67]. However, the word topological was not a buzz word by then. There is, however, an important point to make here. There are four inequivalent L points in the Brillouin zone of these compounds, which means that band inversion occurs at an even number of TRIM of the Brillouin zone [54]. Therefore, according to our discussion of the $\mathbb{Z}_2$ index, there will be no change in sign in the total parity and, therefore, the system before and after band inversion is topologically trivial. The first response in the scientific community when the topological insulators were discovered was to regard these materials as being uninteresting in the new sense of the topologically nontrivial matter [45, 54], even though they were known to host surface states as well [62, 66]. However, just a few years after, a seminal paper in 2011 by Fu [68] proposed that there may exist another category of topological insulators protected by point-group symmetries, which were given the name of *topological crystalline insulators*. Leaded by Fu, in 2012 Hsieh et al. [69] showed experimentally the predictions done by Fu in $Pb_{1-x}Sn_xTe$, in particular, that SnTe was a topological crystalline insulator, whereas PbTe is trivial. In this case, the symmetry involved is mirror symmetry. The $\mathbb{Z}_2$ invariant is zero, but mirror symmetry allows for the introduction of yet another invariant: the *mirror Chern number* [70]. We shall not discuss this invariant in this section, but the idea is as follows: mirror symmetry allows us to separate the Hilbert space into two subspaces characterized by the two eigenvalues of such symmetry. After that, one can compute the Chern number in each of these two subspaces and the difference between the two Chern numbers is the mirror Chern number [49]. The Chern number in general is a topological invariant [11], which is particularly important for time-reversal-symmetry-breaking phases, such as the quantum Hall effect, in which case it is the integer that quantizes the Hall conductance [71]. In time-reversal-symmetry-preserving phases, it is shown to be exactly zero [45]. However, in the case of systems with mirror symmetry, we can make the decomposition that we discussed above and find nonzero Chern numbers for each mirror-symmetry subspace. Their sum, the total Chern number, must still be zero, but their difference (the mirror Chern number) may be nonzero [49, 70, 72].

2.3 Topological Boundaries and Nanoribbons

In this section, we will discuss the two systems that are considered within this Thesis, the most prevalent of which is the topological boundary.

2.3.1 Topological Boundary

A topological boundary is formed when two systems of opposite topological invariant are put together to form a boundary. The bulk-boundary correspondence predicts the existence of edge states (in two-dimensions) or surface states (in three-dimensions). In our case, we can place two insulators of opposite gap in the Dirac low-energy Hamiltonian described above. This is very much like what we did in the SSH chain to find the zero-energy modes. Doing the same in band-inverted junctions was proposed in the 80s in a series of seminal papers by Volkov and Pankratov [62, 63, 66]. For simplicity, we may assume the interface to be a sharp interface, so that we can take the energy gap to be defined as a piece-wise function. This way, we will miss the now called Volkov–Pankratov states [73, 74], which appear when the junction is smoother, but we will still find the topological surface state. All in all, the problem we have to solve is therefore the following

$$\left[v_{\perp}\boldsymbol{\alpha}_{\perp}\cdot\boldsymbol{k}_{\perp}+v_z\alpha_z k_z+\Delta(z)\beta+V_{\mathrm{C}}(z)\right]\boldsymbol{\Psi}(\boldsymbol{r})=E\boldsymbol{\Psi}(\boldsymbol{r})\,, \tag{2.3.1}$$

where $\Delta(z)=E_{\mathrm{G}}(z)/2$. We will take the interface to be at $z=0$. Notice that there are continuous translational and rotational symmetries in the XY-plane. The former implies that $\boldsymbol{k}_{\perp}$ is a good quantum number (we can choose the eigenstates to be eigenstates of the in-plane momentum, i.e. plane waves of the form $\exp\left[\mathrm{i}\,\boldsymbol{k}_{\perp}\cdot\boldsymbol{r}\right]$), which allows us to label the solutions to this problem; the latter implies that the energy (not the state) can only depend on powers of $k_{\perp}=|\boldsymbol{k}_{\perp}|$, but not on k_x, k_y, separately. Therefore, in the following, we will write the bispinor as a function of z only, $\boldsymbol{\Psi}(z)$, remembering that the actual bispinor is actually $\exp(\mathrm{i}\,\boldsymbol{k}_{\perp}\cdot\boldsymbol{r})\boldsymbol{\Psi}(z)$, in virtue of translational symmetry. Since we are taking the interface to be abrupt, $\Delta(z)$ and $V_{\mathrm{C}}(z)$ take the following form

$$f(z)=f_{\mathrm{R}}\Theta(z)+f_{\mathrm{L}}\Theta(-z)\,, \tag{2.3.2}$$

where $f=\Delta$, V_{C}, the subscripts R and L indicate right and left and $\Theta(z)$ is the Heaviside step function. In order to have band inversion, we require that $\Delta_{\mathrm{R}}\Delta_{\mathrm{L}}<0$. For concreteness, we shall take $\Delta_{\mathrm{R}}>0$. We shall refer to the case where $V_{\mathrm{C}}^{\mathrm{R}}=V_{\mathrm{C}}^{\mathrm{L}}$ as the *centered junction* and the case where also $\Delta_{\mathrm{R}}=-\Delta_{\mathrm{L}}$ as the *symmetric junction*. Taking into account that both Δ and V_{C} share the same profile, it is convenient to write them as follows

$$\Delta(z) = \Delta + \lambda \mathrm{sgn}(z) , \qquad V_{\mathrm{C}}(z) = V_0 + \gamma \Delta(z) , \tag{2.3.3}$$

where we have introduced

$$\Delta = \frac{1}{2}(\Delta_{\mathrm{R}} + \Delta_{\mathrm{L}}) , \qquad \lambda = \frac{1}{2}(\Delta_{\mathrm{R}} - \Delta_{\mathrm{L}}) , \tag{2.3.4a}$$

$$\gamma = \frac{V_{\mathrm{C}}^{\mathrm{R}} - V_{\mathrm{C}}^{\mathrm{L}}}{2\lambda} , \qquad V_0 = V_{\mathrm{C}}^{\mathrm{L}} - \gamma \Delta_{\mathrm{L}} = V_{\mathrm{C}}^{\mathrm{R}} - \gamma \Delta_{\mathrm{R}} . \tag{2.3.4b}$$

Let us introduce the following quantities

$$d = \frac{v_z}{\lambda} , \quad \xi = \frac{z}{d} , \quad \boldsymbol{\kappa} = \frac{v_\perp}{v_z} d\, \boldsymbol{k}_\perp , \quad \delta = \frac{\Delta}{\lambda} , \quad \varepsilon = \frac{E - V_0}{\lambda} . \tag{2.3.5}$$

Then, we can write the Dirac equation as $\mathcal{H}\boldsymbol{\Psi}(\xi) = (\varepsilon - \gamma\delta)\, \boldsymbol{\Psi}(\xi)$, where

$$\mathcal{H} = \mathcal{H}_0 + (\beta + \gamma)\, \mathrm{sgn}(\xi) , \tag{2.3.6}$$

and

$$\mathcal{H}_0 = -\mathrm{i}\, \alpha_z \partial_\xi + \boldsymbol{\alpha}_\perp \cdot \boldsymbol{\kappa} + \delta\beta . \tag{2.3.7}$$

with $\partial_\xi = d/d\xi$. We have not absorbed the term $\gamma\delta$ in ε on purpose, as we shall see that it simplifies the final results. As is customary with the Dirac equation and as we did in the SSH chain, we will square the Hamiltonian. Taking into account that $\mathcal{H}\boldsymbol{\Psi}(\xi) = (\varepsilon - \gamma\delta)\, \boldsymbol{\Psi}(\xi)$, we obtain

$$\left[-\partial_\xi^2 + \pi^2(\xi) + U(\xi)\right] \boldsymbol{\Psi}(\xi) = 0 , \tag{2.3.8}$$

where

$$\pi^2(\xi) = \kappa^2 + \left[\delta + \mathrm{sgn}(\xi)\right]^2 - \left[\varepsilon - \gamma \mathrm{sgn}(\xi) - \gamma\delta\right]^2 , \tag{2.3.9}$$

and

$$U(\xi) = -2\mathrm{i}\, \alpha_z(\beta + \gamma)\delta(\xi) . \tag{2.3.10}$$

From the form of $\pi^2(\xi)$, it is now clear that the energy can only depend on powers of $k_\perp$, as was already apparent from rotational symmetry. We can solve Eq. (2.3.8) on both sides of $\xi = 0$, where $U(\xi) = 0$ and $\pi(\xi)$ is a constant. Doing so, we obtain

$$\boldsymbol{\Psi}(\xi) = \exp\left[-\pi(\xi)|\xi|\right] \boldsymbol{\Phi} , \tag{2.3.11}$$

where we have assumed $\pi(\xi) > 0$ so that we can find bound-state solutions and we have already exploited the fact that $\boldsymbol{\Psi}(\xi)$ has to be continuous at $\xi = 0$, which allows us to set the same constant vector $\boldsymbol{\Phi}$ on both sides. By integrating Eq. (2.3.8) around $\xi = 0$, we obtain a discontinuity on the derivative of the bispinor

$$-\partial_\xi \boldsymbol{\Psi}(\xi)|_{\xi=0^+} + \partial_\xi \boldsymbol{\Psi}(\xi)|_{\xi=0^-} = 2\mathrm{i}\,\alpha_z\,(\beta+\gamma)\,\boldsymbol{\Psi}(0)\,. \tag{2.3.12}$$

Taking into account the form of $\boldsymbol{\Psi}(\xi)$, this equation can be written as follows

$$2\mathrm{i}\,\alpha_z\,(\beta+\gamma)\,\boldsymbol{\Phi} = \left[\pi^+ + \pi^-\right]\boldsymbol{\Phi}\,, \tag{2.3.13}$$

with $\pi^\pm \equiv \pi(0^\pm)$. This equation can be interpreted to be an eigenvalue problem. The four eigenvalues of the matrix on the left are $\pm 2\sqrt{1-\gamma^2}$, each of which is doubly degenerate. However, since we have chosen $\pi(\xi)$ to be positive, then we have to disregard the two negative eigenvalues. Also notice that we have chosen $\pi(\xi)$ to be real, which implies that $|\gamma| < 1$ or, equivalently, $|\Delta_\mathrm{R} - \Delta_\mathrm{L}| > |V_\mathrm{L} - V_\mathrm{R}|$. That is, the gaps on either side of the junction must overlap in order to have bound states. All in all, we obtain that

$$\pi^+ + \pi^- = 2\sqrt{1-\gamma^2}\,, \tag{2.3.14}$$

with double degeneracy. This equation can be solved for the energies to give

$$\varepsilon_\pm = \pm\sqrt{1-\gamma^2}\,\kappa\,. \tag{2.3.15}$$

This is nothing but a Dirac cone, very much like those that appear in bulk graphene. As expected from the bulk-boundary correspondence, we have a topological surface state at the boundary. The fact that the dispersion is a Dirac cone can be understood from the fact that there is time-reversal symmetry, as we shall see in a moment. The two normalized eigenvectors corresponding to the two positive eigenvalues are

$$\boldsymbol{\Phi}_a = \frac{1}{\sqrt{2}}\begin{pmatrix} -\mathrm{i}\sqrt{1-\gamma} \\ 0 \\ \sqrt{1+\gamma} \\ 0 \end{pmatrix}\,, \qquad \boldsymbol{\Phi}_b = \frac{1}{\sqrt{2}}\begin{pmatrix} 0 \\ \mathrm{i}\sqrt{1-\gamma} \\ 0 \\ \sqrt{1+\gamma} \end{pmatrix}\,. \tag{2.3.16}$$

These two eigenvectors are related by time-reversal symmetry. Indeed, if we apply $\mathcal{T} = \left(\tau_0 \otimes \mathrm{i}\,\sigma_y\right)\mathcal{K}$ to $\boldsymbol{\Phi}_a$ we obtain $\boldsymbol{\Phi}_b$ and vice versa. That is, $\boldsymbol{\Phi}_b = \mathcal{T}\boldsymbol{\Phi}_a$. Therefore, $\boldsymbol{\Phi}_a$ and $\boldsymbol{\Phi}_b$ form a Kramers' pair. Notice that these two eigenvectors are $\boldsymbol{k}$-independent. As a consequence, we see that $\boldsymbol{\Psi}(\xi) = \exp\left[-\pi(\xi)|\xi|\right]\boldsymbol{\Phi}_\alpha$, with $\alpha = a, b$, does not satisfy the Dirac equation (this can be seen by mere substitution into the equation), unless $\boldsymbol{\kappa} = 0$. This is what we would expect from time-reversal symmetry, since the only time-reversal-symmetric-momentum of this model is $\boldsymbol{\kappa} = 0$. Therefore, we only expect Kramers' pairs to exist at $\boldsymbol{\kappa} = 0$. Away from $\boldsymbol{\kappa} = 0$, the degeneracy breaks down and the two states split apart. As a result, away from $\boldsymbol{\kappa} = 0$, the vector in $\boldsymbol{\Psi}(\xi)$ is no longer $\boldsymbol{\Phi}_a$ or $\boldsymbol{\Phi}_b$, but rather a linear combination of the two. Let us then rename $\boldsymbol{\Phi}_a$ as $\boldsymbol{\Phi}_0$ and write $\boldsymbol{\Phi}_b$ simply as $\mathcal{T}\boldsymbol{\Phi}_0$. Then, away from $\boldsymbol{\kappa} = 0$, the eigenvectors will be given by

$$\boldsymbol{\Phi}_\kappa = a\,\boldsymbol{\Phi}_0 + b\,\mathcal{T}\boldsymbol{\Phi}_0\,. \tag{2.3.17}$$

These of course still satisfy Eq. (2.3.13). Let us consider the symmetric junction for simplicity, in which case $\pi(\xi) = 1$. If we apply the Dirac Hamiltonian to $\boldsymbol{\Psi}(\xi)$ taking as constant vector $\boldsymbol{\Phi}_{\boldsymbol{\kappa}}$, we are led to the following equation

$$[\boldsymbol{\alpha}_{\perp} \cdot \boldsymbol{\kappa}]\, \boldsymbol{\Phi}_{\boldsymbol{\kappa}} = \varepsilon \boldsymbol{\Phi}_{\boldsymbol{\kappa}} \,. \tag{2.3.18}$$

In order to obtain the coefficients a and b, we can project on the left with $\boldsymbol{\Phi}_0$ and $\mathcal{T}\boldsymbol{\Phi}_0$, which leads to

$$\begin{pmatrix} \boldsymbol{\Phi}_0^{\dagger}\, [\boldsymbol{\alpha}_{\perp} \cdot \boldsymbol{\kappa}]\, \boldsymbol{\Phi}_0 & \boldsymbol{\Phi}_0^{\dagger}\, [\boldsymbol{\alpha}_{\perp} \cdot \boldsymbol{\kappa}]\, \mathcal{T}\boldsymbol{\Phi}_0 \\ (\mathcal{T}\boldsymbol{\Phi}_0)^{\dagger}\, [\boldsymbol{\alpha}_{\perp} \cdot \boldsymbol{\kappa}]\, \boldsymbol{\Phi}_0 & (\mathcal{T}\boldsymbol{\Phi}_0)^{\dagger}\, [\boldsymbol{\alpha}_{\perp} \cdot \boldsymbol{\kappa}]\, \mathcal{T}\boldsymbol{\Phi}_0 \end{pmatrix} \begin{pmatrix} a \\ b \end{pmatrix} = \varepsilon \begin{pmatrix} a \\ b \end{pmatrix} . \tag{2.3.19}$$

We can interpret this equation as though we were applying first order degenerate perturbation theory. If the off-diagonal terms are non-zero, this means that the Kramers' pair mix and there is a degeneracy breaking. This equation can also be interpreted as the Hamiltonian in the subspace of the surface states, which is why the matrix on the left is commonly referred to as the *Hamiltonian for the surface states* [53]. If we perform the calculations, the resulting matrix is given by

$$\mathcal{H}_S = \kappa_y \sigma_x - \kappa_x \sigma_y = (\boldsymbol{\sigma} \times \boldsymbol{\kappa})_z \,, \tag{2.3.20}$$

with eigenvalues the two Dirac cones (positive and negative) that we obtained earlier and we obtain that $b = \mp \mathrm{i}\, a \exp{(\mathrm{i}\, \theta_{\boldsymbol{\kappa}})}$, where the upper sign corresponds to the upper cone, $\varepsilon_+ = \kappa$, and the lower sign corresponds to the lower cone, $\varepsilon_- = -\kappa$. Here, $\theta_{\boldsymbol{\kappa}} = \arg\left(\kappa_x + \mathrm{i}\, \kappa_y\right)$. From normalization, we obtain that $|a|^2 = 1/2$, which leaves an arbitrary phase to a. We will choose it to be $\exp(-\mathrm{i}\, \theta_{\boldsymbol{\kappa}}/2)$. As a result, the two eigenstates of $\mathcal{H}_S$ can be written as follows

$$\boldsymbol{\chi}_s = \mathcal{R}_z\, (\theta_{\boldsymbol{\kappa}})\, \boldsymbol{\chi}_s^0 \,, \tag{2.3.21}$$

where $s = \pm$ and

$$\mathcal{R}_z\, (\theta_{\boldsymbol{\kappa}}) = \exp\left(-\mathrm{i}\, \frac{\theta_{\boldsymbol{\kappa}}}{2} \sigma_z\right) , \quad \boldsymbol{\chi}_s^0 = \begin{pmatrix} 1 \\ -s\mathrm{i} \end{pmatrix} . \tag{2.3.22}$$

This result is very much like the one we obtained in the SSH chain [cf. Eq. (2.1.26)]. It means that upon a 2π rotation in $\boldsymbol{\kappa}$-space, the state acquires a phase of π, showing the non-trivial topology of these states. It is also important to note that the Hamiltonian for the surface states is a Rashba Hamiltonian without quadratic terms. The consequence of having such a Hamiltonian is that the surface states are *spin-momentum-locked* and the states of the two cones have opposite *helicities* [20]. In order to see this, we can calculate the spin texture by evaluating $\langle \boldsymbol{\sigma} \rangle$, taking into account the form of the states in Eq. (2.3.21). If we do so, the result is the following

$$\langle \boldsymbol{\sigma} \rangle = \pm\, (\sin \theta_{\boldsymbol{\kappa}}, -\cos \theta_{\boldsymbol{\kappa}}, 0) \,, \tag{2.3.23}$$

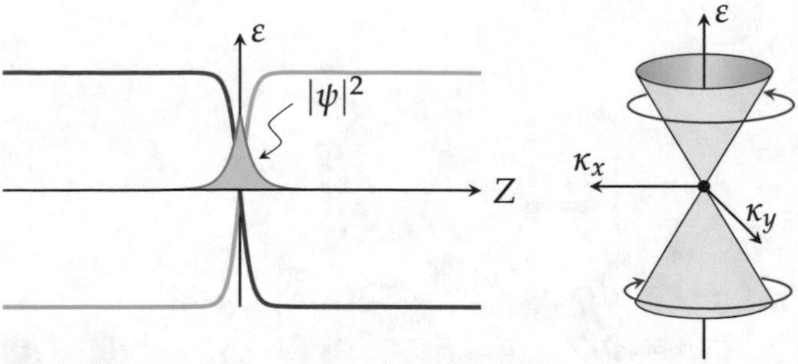

Fig. 2.11 **Topological surface state** localized at the interface where band inversion occurs. The localized state on the left corresponds the Dirac point on the right, marked by a black dot at the vertex of the cones. The Dirac point is protected by time-reversal symmetry. Away from it, Kramers' degeneracy no longer holds and the states at the Dirac point split apart, leading to the Dirac cones. These cones have opposite helicities, which are perpendicular to momentum at all times, as schematically indicated by the curved arrows

which implies that $\langle \boldsymbol{\sigma} \rangle$ and $\boldsymbol{\kappa}$ are perpendicular to each other, which corresponds to the feature of having spin-momentum-locking, sometimes referred to as *helical spin polarization* [49]. The $\pm$ signs correspond to the two opposite Dirac cones, which implies that they have opposite helicities. The main implication is that back-scattering from $\boldsymbol{\kappa}$ to $-\boldsymbol{\kappa}$ is forbidden (if a non-spin dependent perturbation is applied), since that would oblige to change sign of $\langle \boldsymbol{\sigma} \rangle$.

To conclude with the topological boundary, let us summarize what we have learnt about the surface states of topological insulators. Time-reversal symmetry requires, by virtue of Kramers' theorem, to have a pair of degenerate states at $\boldsymbol{\kappa} = 0$. Away from $\boldsymbol{\kappa} = 0$, Kramers' theorem does not hold anymore and the two Kramers' pairs hybridize due to spin-orbit coupling, splitting to form the two Dirac cones. Finally, these surface states have the feature of being helical, meaning that spin is always perpendicular to momentum. Moreover, the upper and lower cones have opposite helicities. This summary is schematically displayed in Fig. 2.11. It is important to make a final remark. Although we have not stated this explicitly, it is important to make a distinction upon where the interface is with respect to the quintuple layers. In our case, we implicitly considered the surface parallel to the quintuple layers, which corresponds to the (111) surface [cf. Fig. 2.9]. The results for the surface states change if other directions are chosen [20], but there is always an (elliptical) Dirac cone and spin-momentum-locking.

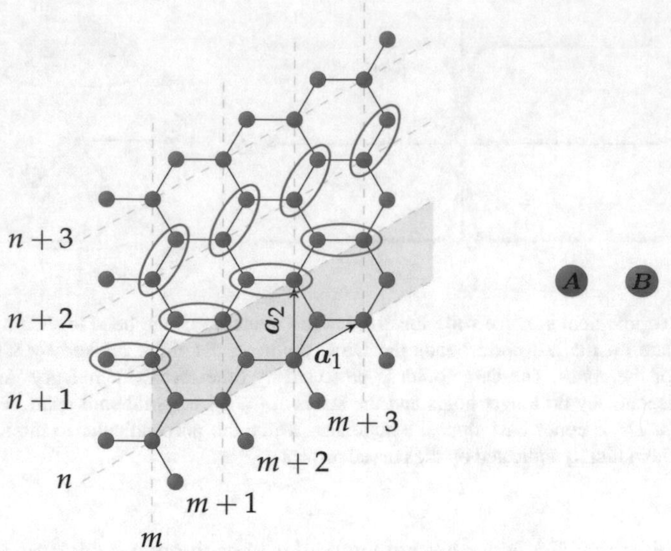

Fig. 2.12 Zigzag and bearded nanoribbons. The vertical direction is taken to be periodic and the transverse direction is finite. The left-hand side represents a bearded termination, whereas the right-hand side represents a zigzag termination. Ovals show ways to arrange the red and blue atoms depending on the desired edges, the red and blue ovals corresponding to bearded and zigzag terminations, respectively. A supercell is shown in yellow. The lattice vectors are $\boldsymbol{a}_1$ and $\boldsymbol{a}_2$

2.3.2 Graphene Nanoribbons

Let us turn our attention to graphene nanoribbons, where the system is periodic in only one of the two directions. There are several types of nanoribbons, depending on their termination. We will be interested in *zigzag*, *bearded* and *armchair* nanoribbons.

Zigzag and Bearded Nanoribbons

Let us begin with *zigzag* and *bearded* nanoribbons. We shall choose the lattice with the orientation shown in Fig. 2.12. The system is taken to be periodic along the vertical direction. Due to translational symmetry, there is a conserved momentum along that direction, k. In the transverse direction, the momentum is not a good quantum number anymore. Rather, it gets quantized, although not in a straightforward manner, as we shall see. Since the lattice constant is still that of bulk graphene, $a = 1$, then k is restricted to live within $k \in [-\pi, \pi)$. Let us consider a first-nearest-neighbour tight-binding model for the zigzag nanoribbon. For that matter, let us introduce the following two vectors

$$\boldsymbol{a}_1 = \frac{1}{2}\left(1, \sqrt{3}\right), \qquad \boldsymbol{a}_2 = (0, 1). \tag{2.3.24}$$

Linear combinations of these with integer coefficients allow to build the lattice. However, one has to take care with the fact that the lattice is finite in the transverse direction

$$\boldsymbol{R}_{m,n} = m\boldsymbol{a}_1 + n\boldsymbol{a}_2 \,, \qquad m = 1, \ldots, N_x \,, \qquad n \in \mathbb{Z} \,. \tag{2.3.25}$$

The two atoms in the ovals of Fig. 2.12 are linked by the vectors

$$\boldsymbol{\delta}_z = \frac{1}{2\sqrt{3}}\left(1, \sqrt{3}\right) \,, \qquad \boldsymbol{\delta}_b = -\frac{1}{\sqrt{3}}(1, 0) \,, \tag{2.3.26}$$

where z stands for zigzag and b for bearded. The position of the A atoms (blue in the figure) will be given by $\boldsymbol{R}^A = \boldsymbol{R}_{m,n}$ and those of the B atoms (red) by $\boldsymbol{R}^B = \boldsymbol{R}_{m,n} + \boldsymbol{\delta}_{z/b}$. We will then say that an atom is at position (m, n) and it is of type A, B. In the zigzag nanoribbon, an A atom in position (m, n) is connected to the B atoms at (m, n), $(m, n-1)$, $(m-1, n)$, whereas in the bearded nanoribbon an A atom in (m, n) is connected to the B atoms at (m, n), $(m+1, n)$ and $(m+1, n-1)$. By hopping from A in (m, n) to B in $(m, n+q)$, with q an integer, we accumulate a phase factor of $\exp(\mathrm{i}\,qk)$, since the process amounts to translation by q-units in the vertical direction. On the other hand, hopping from A in (m, n) to B in $(m+q, n)$ does not involve the inclusion of phase factors, since we are within the same supercell. Therefore, a simple tight-binding model for the zigzag nanoribbon can be written as follows ($t = 1$)

$$\left(1 + e^{-\mathrm{i}k}\right)\psi_B(m) + \psi_B(m-1) = E\psi_A(m) \,, \tag{2.3.27a}$$

$$\psi_A(m+1) + \left(1 + e^{\mathrm{i}k}\right)\psi_A(m) = E\psi_B(m) \,, \tag{2.3.27b}$$

and that of a bearded nanoribbon can be written as

$$\psi_B(m) + \left(1 + e^{-\mathrm{i}k}\right)\psi_B(m+1) = E\psi_A(m) \,, \tag{2.3.28a}$$

$$\psi_A(m) + \left(1 + e^{\mathrm{i}k}\right)\psi_A(m-1) = E\psi_B(m) \,, \tag{2.3.28b}$$

If we had bulk graphene, we would be allowed to introduce a conserved transverse momentum as well, and would have $\psi_\alpha(m) \sim \exp\left(\mathrm{i}\,m\boldsymbol{k}\cdot\boldsymbol{a}_1\right)$. In turn, from the zigzag and bearded nanoribbons' equations we would obtain

$$\begin{pmatrix} 0 & \Delta_{z/b}(\boldsymbol{k}) \\ \Delta^*_{z/b}(\boldsymbol{k}) & 0 \end{pmatrix}\begin{pmatrix} A \\ B \end{pmatrix} = E\begin{pmatrix} A \\ B \end{pmatrix} . \tag{2.3.29}$$

where

$$\Delta_z(\boldsymbol{k}) = 1 + e^{-\mathrm{i}k_1} + e^{-\mathrm{i}k_2} \,, \qquad \Delta_b(\boldsymbol{k}) = 1 + e^{\mathrm{i}k_1} + e^{\mathrm{i}(k_1-k_2)} \,, \tag{2.3.30}$$

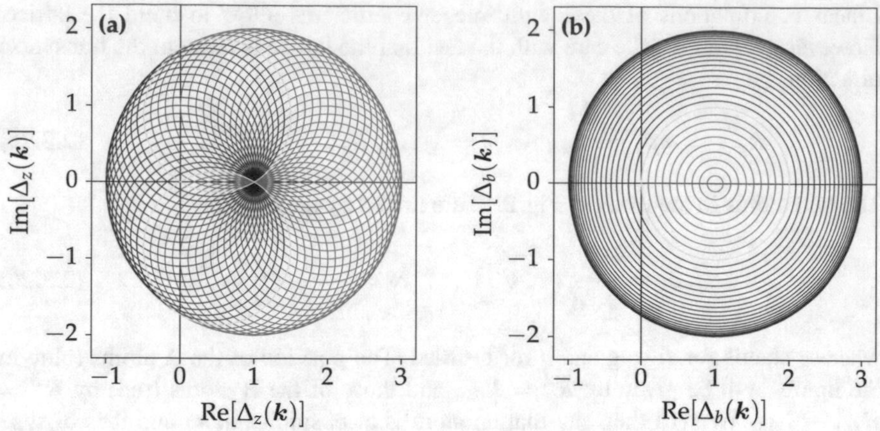

Fig. 2.13 Topology of zigzag and bearded nanoribbons. **a** $\Delta_z(\boldsymbol{k})$ and **b** $\Delta_b(\boldsymbol{k})$ in the complex plane. Each of the circles corresponds to a fixed value of k_2, as k_1 is varied from 0 to 2π. Yellow circles touch the origin and correspond to $k_2 = \pm 2\pi/3$. The teal-coloured circles correspond to **a** $k_2 < |2\pi/3|$ and **b** $k_2 > |2\pi/3|$ and do not enclose the origin, whereas the dark red circles enclose the origin and correspond to **a** $k_2 > |2\pi/3|$ and **b** $k_2 < |2\pi/3|$ [42]

with $k_i = \boldsymbol{k} \cdot \boldsymbol{a}_i$. Obviously, both equations lead to the same dispersion, that is, $|\Delta_z(\boldsymbol{k})| = |\Delta_b(\boldsymbol{k})|$. However, we already know the importance of studying $\Delta(\boldsymbol{k})$ as $\boldsymbol{k}$ is evolved throughout the Brillouin zone. Since we will be interested in finite systems afterwards, it is convenient to think about the evolution of $\Delta_\alpha(\boldsymbol{k})$ as k_1 is varied from 0 to 2π, keeping $k_2 \in [-\pi, \pi)$ fixed. For $\Delta_z(\boldsymbol{k})$, it is then clear that for each value of k_2 we have a circle of radius $R = 1$, centered at $C(k_2) = 1 + \exp(-\mathrm{i}\, k_2)$. We have then three options: the circle encloses the origin ($|C(k_2)| < R$), it goes straight through it ($|C(k_2)| = R$) or it does not enclose it ($|C(k_2)| > R$). By looking at $C(k_2)$, we can see that $k_2 = \pm 2\pi/3$ corresponds to having the two corresponding circles going through the origin. Since we are looking at circles, these can only go through the origin once each. Touching the origin implies that $\Delta_z(\boldsymbol{k}) = 0$. Therefore, these two touching points are the two Dirac points of the Brillouin zone, K and K'. If $|k_2| > 2\pi/3$, then $|C(k_2)| < R$ and the circles enclose the origin and otherwise if $|k_2| < 2\pi/3$. We shall see the importance of these results shortly, which are shown in Fig. 2.13a. In the case of $\Delta_b(\boldsymbol{k})$, we can observe that the curves correspond once again to circles but this time they have a fixed center, $C = 1$, and a radius $R(k_2) = |1 + \exp(-\mathrm{i}\, k_2)|$. Therefore, the radius and the center have swapped places with respect to the zigzag nanoribbon and the arguments we just presented are inverted. That is, the circles enclose the origin if $|k_2| < 2\pi/3$ and they do not if $|k_2| > 2\pi/3$. The results are shown in Fig. 2.13b.

Let us turn our attention to the equations for the finite systems. If we assume that $E \neq 0$, from both sets of equations [cf. Eqs. (2.3.27), (2.3.28)] we find

$$\left[|f(k)|^2+1\right]\psi_A(m)+f(k)\psi_A(m+1)+f^*(k)\psi_A(m-1)=E^2\psi_A(m)\,, \tag{2.3.31}$$

and $\psi_B(m)$ can be obtained from

$$\psi_B(m)=\frac{1}{E}\left[\psi_A(m+1)+f^*(k)\psi_A(m)\right]\,, \tag{2.3.32}$$

for the zigzag nanoribbon and

$$\psi_B(m)=\frac{1}{E}\left[\psi_A(m)+f^*(k)\psi_A(m-1)\right]\,, \tag{2.3.33}$$

for the bearded nanoribbon. Here we have introduced $f(k)=1+\exp(-ik)$. It is important to notice the fact that Eq. (2.3.31) is the same for both types of edges, which implies that the energies resulting from this equation cannot be edge states, since its only the bulk of zigzag and bearded nanoribbons that is the same. From Eq. (2.3.31) we also see that the spectrum will be symmetric around $E=0$. Indeed, the spectrum corresponding to this equation for a large nanoribbon of $N_x=200$ corresponds to the khaki lines in Fig. 2.14a. A bulk state is shown in green in Fig. 2.14b, where it can be seen that the probability density spreads throughout the ribbon and is zero at the edges.

Let us now explore the case where $E=0$. In that case, the equations for the zigzag and bearded nanoribbons [cf. Eqs. (2.3.27), (2.3.28)] imply that

$$\begin{aligned}|\psi_B(m+1)|^2&=|f(k)|^{-2m}|\psi_B(1)|^2\,,\\ |\psi_A(m+1)|^2&=|f(k)|^{2m}|\psi_A(1)|^2\,.\end{aligned} \tag{2.3.34}$$

If $|f(k)|<1$, $|\psi_B(m+1)|^2>|\psi_B(1)|^2$ and $|\psi_A(m+1)|^2<|\psi_A(1)|^2$. The opposite happens with $|f(k)|>1$. Which one we choose, either $|f(k)|>1$ or $|f(k)|<1$, depends on the boundary conditions. A zigzag nanoribbon requires that [see Fig. 2.12] $\psi_B(0)=\psi_A(N_x+1)=0$, which implies that we must choose $|f(k)|<1$, since it implies that $|\psi_B|^2$ is exponentially localized on the right and $|\psi_A|^2$ is on the left. On the other hand, a bearded nanoribbon requires that $\psi_A(0)=0$ and $\psi_B(N_x+1)=0$, so we must choose $|f(k)|>1$ in that case. The region where $|f(k)|<1$ and $|f(k)|>1$ correspond, respectively, to $k>|2\pi/3|$ and $k<|2\pi/3|$. In Fig. 2.14a, we have plotted in dark blue the zero energy states corresponding to the bearded nanoribbon and in light coral those corresponding to the zigzag nanoribbon. At the point where $|f(k)|=1$, the edge states become degenerate with the bulk, as is apparent from that figure.

It is important to mention that if the ribbon is sufficiently narrow, then the edge states can hybridize with the bulk states that are closest in energy, that is, those around $k=|2\pi/3|$, and those states will gap out. However, a little further away from $k=|2\pi/3|$, the states are sufficiently far from bulk states and they are zero energy modes. This is why in Fig. 2.14a we have chosen a ribbon of $N_x=200$ sites. In Fig. 2.14b we show two edge states for the zigzag nanoribbon, left- and right-

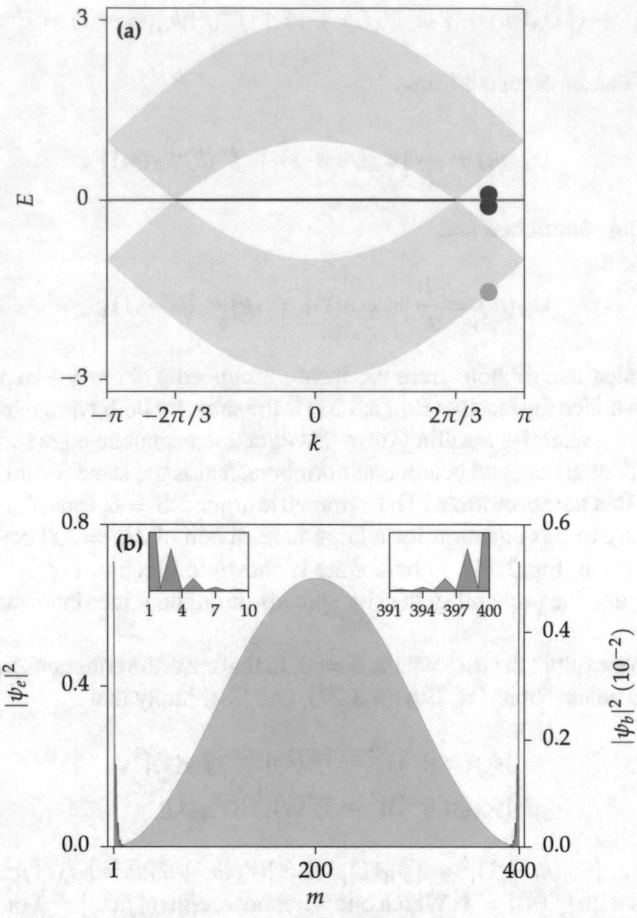

Fig. 2.14 Dispersion, edge and bulk states of zigzag and bearded nanoribbons. a Dispersion for a nanoribbon of $N_x = 200$, where bulk states are shown in khaki and coincide for both kinds of ribbons. Zero energy dark blue and light coral lines correspond to edge states in bearded and zigzag nanoribbons, respectively. Filled circles correspond to the states that are plotted in **b** with the probability densities for the bulk state in green, $|\psi_b|^2$, and the edge states in blue and red, $|\psi_e|^2$. The index m indicates the site, odd and even numbers corresponding to A and B atoms, respectively. Insets show how the edge states have all their weight in either one of the two atoms (A for the left edge, B for the right edge)

hand sides, corresponding to the blue and red dots in Fig. 2.14a, respectively. As we can see in the inset, the probability takes only non-zero values on the A sites (odd numbered in the figure) for the left-hand side of the ribbon and on the B sites (even numbered) for the right-hand side. This explains the flatness of the bands: indeed, since the weight on atoms of the opposite sublattice is exactly zero, there is no way for tunneling into those sites and spreading into the bulk, so there can be no dispersion. It is possible to understand the presence of these edge states from

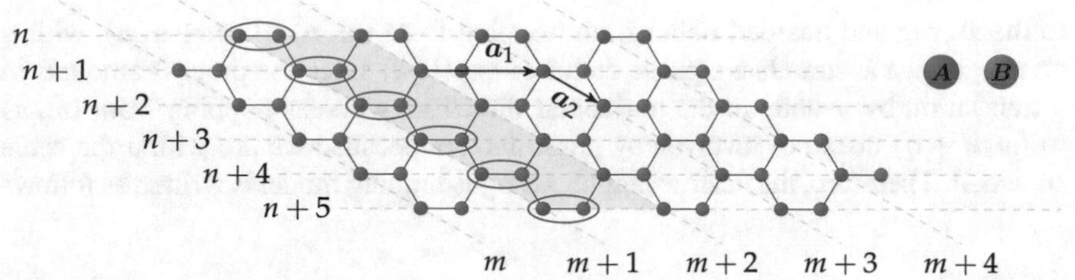

Fig. 2.15 **Armchair nanoribbon**. The longitudinal direction is taken to be periodic and the transverse direction is finite. Ovals show a way to arrange the *A* and *B* atoms. A supercell is shown in yellow. The lattice vectors are $\boldsymbol{a}_1$ and $\boldsymbol{a}_2$

topology arguments. Indeed, we have just obtained that the zigzag nanoribbon hosts edge states if $|f(k)| < 1$, whereas the bearded nanoribbon does if $|f(k)| > 1$. In our analysis of the bulk Hamiltonian when we have seen that $\Delta_{z,b}(\boldsymbol{k})$ describes circles enclosing the origin precisely when $|f(k)| < 1$ for zigzag nanoribbons and likewise for bearded nanoribbons if $|f(k)| > 1$. Therefore, the existence of these edge states stems from a *bulk-boundary correspondence* [42, 75, 76], very much like in the SSH chain and the topological boundary.

Armchair Nanoribbons

Let us now turn our attention to *armchair nanoribbons*, as shown in Fig. 2.15, where the horizontal direction is taken to be periodic and the transverse direction is finite. Notice that the lattice constant is now $\sqrt{3}$ times larger than that of bulk graphene, so the conserved momentum along the horizontal direction will be restricted to a Brillouin zone that is $\sqrt{3}$ times smaller. In any case, we will take here the lattice constant to be equal to 1, so that the bulk lattice constant is $1/\sqrt{3}$. The lattice is then generated by the two vectors

$$\boldsymbol{a}_1 = (1, 0)\,, \qquad \boldsymbol{a}_2 = \frac{1}{2\sqrt{3}}\left(\sqrt{3}, -1\right)\,, \tag{2.3.35}$$

so that a position in the lattice is given by

$$\boldsymbol{R}_{m,n} = m\boldsymbol{a}_1 + n\boldsymbol{a}_2\,, \qquad m \in \mathbb{Z}\,, \qquad n = 1, \ldots, N_y\,. \tag{2.3.36}$$

The *A* and *B* atoms inside an oval (see Fig. 2.15) are linked by a vector $\boldsymbol{\delta} = (1/3, 0)$. This way, the position of the *A* atoms is $\boldsymbol{R}^A = \boldsymbol{R}_{m,n}$ and that of the *B* atoms is $\boldsymbol{R}^B = \boldsymbol{R}_{m,n} + \boldsymbol{\delta}$. Just like with the zigzag and bearded lattices, we will say that an atom is at position (m, n) and is of type A, B.

Since the lattice is periodic along the horizontal direction, there is a good quantum number along that direction, k, restricted to $k \in [-\pi, \pi)$. Recall that we are expressing everything in units of the lattice constant for this nanoribbon, which is $\sqrt{3}$ times larger than that of bulk graphene. In the figure, we can see that an *A* atom in (m, n) is connected to a *B* atom in (m, n), $(m, n-1)$ and $(m-1, n+1)$. Similarly to the zigzag and bearded ribbons, on hopping from (m, n) to $(m+q, n)$, with q an integer, we accumulate a phase factor of $\exp(\mathrm{i}\, qk)$ since the process amounts to a translation by q-units in the horizontal direction, whereas hopping from (m, n) to $(m, n+q)$ does not involve any phase factors because we are within the same supercell. Therefore, the nearest-neighbour tight-binding model is written as follows

$$\psi_B(n) + \psi_B(n-1) + e^{-\mathrm{i}k}\psi_B(n+1) = E\psi_A(n) \tag{2.3.37a}$$

$$\psi_A(n) + \psi_A(n+1) + e^{\mathrm{i}k}\psi_A(n-1) = E\psi_B(n)\,. \tag{2.3.37b}$$

If we do the same as with the other ribbons, in the bulk we would have $\psi_\alpha(n) \propto \exp(\mathrm{i}\, n\boldsymbol{k}\cdot\boldsymbol{a}_2)$ and the relevant quantity to look at would be

$$\Delta_a(\boldsymbol{k}) = 1 + e^{-\mathrm{i}k_2} + e^{-\mathrm{i}(k_1-k_2)}\,, \tag{2.3.38}$$

where again $k_1 = \boldsymbol{k}\cdot\boldsymbol{a}_1$ and $k_2 = \boldsymbol{k}\cdot\boldsymbol{a}_2$. Following the same procedure as with the other nanoribbons, we want to see the evolution of $\Delta_a(\boldsymbol{k})$ in the complex plane as we vary $k_2 \in [0, 2\pi)$ while keeping k_1 fixed since now k_1 is the one that will be a good quantum number in the nanoribbon. In order to visualize what would happen, it is convenient to write $\Delta_a(\boldsymbol{k})$ as follows

$$\Delta_a(\boldsymbol{k}) = 1 + 2e^{-\mathrm{i}k_1/2}\cos\left(\frac{k_1}{2} - k_2\right). \tag{2.3.39}$$

Consider the case where $k_1 = 0$. In that case, there is only a real part and, as we vary $k_2 \in [0, 2\pi)$, Δ_a will take values from -1 to 3. Hence, if $k_1 = 0$ we have a segment that passes through the origin and 1. This segment is actually an elliptical loop of major axis equal to 2 and minor axis of size 0, which resembles a segment of size 4 when plotted. Therefore, it passes through the origin twice, when $k_2 = \pi/2$ and $3\pi/2$. Now, let us consider the case where $k_1 = \pm\pi$. In that case, the real part is 1 and the imaginary part is $\mp 2\cos(k_1/2 - k_2)$. We therefore have a vertical segment (loop) that passes through 1 and extends from $-2\,\mathrm{i}$ to $2\,\mathrm{i}$. All other cases where $k_1 \neq 0$ or π can be obtained just by tilting the elliptic loop by an angle of $-k_1/2$, measured from the horizontal axis to the major axis of the ellipse. Hence, we see that, in contrast to the other two ribbons, there are no loops enclosing the origin, except for the loop that touches it twice, the loop at $k_1 = 0$, which correspond to the two Dirac points in bulk graphene. As we will see, there will be no edge states in consequence of the bulk-boundary correspondence. This discussion is summarized in Fig. 2.16.

Let us then turn our attention to the finite ribbon. This model can be solved by performing a gauge transformation such that

$$\psi_A(n) \to \exp\left(\mathrm{i}\, k\frac{n-1}{2}\right)\psi_A(n)\,, \qquad \psi_B(n) \to \exp\left(\mathrm{i}\, k\frac{n}{2}\right)\psi_B(n)\,, \tag{2.3.40}$$

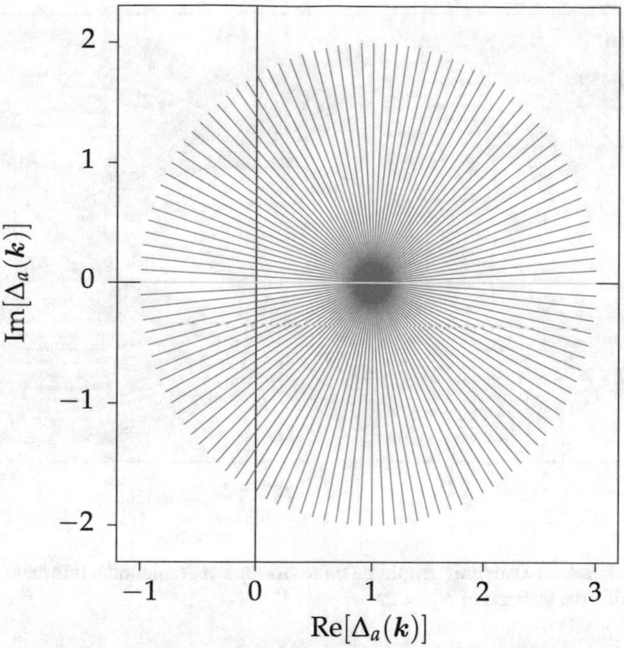

Fig. 2.16 Topology of armchair nanoribbons. $\Delta_a(\boldsymbol{k})$ in the complex plane. Each of the circles corresponds to a fixed value of k_1, as k_2 is varied from 0 to 2π. Lines here are actually elliptic loops of minor axis of size zero. That is, they are not open paths but loops. The yellow lines touch the origin and correspond to $k_1 = 0$, which occurs twice due to the looped nature of these lines. The teal-coloured ellipses correspond to $k_1 \neq 0$, non of which can enclose the origin due to having zero minor axis

which transforms the set of Eqs. (2.3.37) to

$$e^{\mathrm{i}k/2}\psi_B(n) + \psi_B(n-1) + \psi_B(n+1) = E\psi_A(n) \tag{2.3.41a}$$

$$e^{-\mathrm{i}k/2}\psi_A(n) + \psi_A(n+1) + \psi_A(n-1) = E\psi_B(n)\,. \tag{2.3.41b}$$

The boundary conditions require that the amplitudes $\psi_A(n)$ and $\psi_B(n)$ both vanish at $n = 0$ and $n = N_y + 1$ [see Fig. 2.15]. Notice that the gauge transformed amplitudes satisfy exactly the same requirement. If we use the following ansatz

$$\psi_A(n) = A_+e^{\mathrm{i}qn} + A_-e^{-\mathrm{i}qn}\,, \qquad \psi_B(n) = B_+e^{\mathrm{i}qn} + B_-e^{-\mathrm{i}qn}\,, \tag{2.3.42}$$

the boundary conditions at $n = 0$ imply that $A_+ = -A_- \equiv A$ and $B_+ = -B_- \equiv B$. Therefore

$$\psi_A(n) = 2\mathrm{i}\,A\sin(qn)\,, \qquad \psi_B(n) = 2\mathrm{i}\,B\sin(qn)\,. \tag{2.3.43}$$

The boundary condition at $N_y + 1$ implies on the other hand the quantization of q

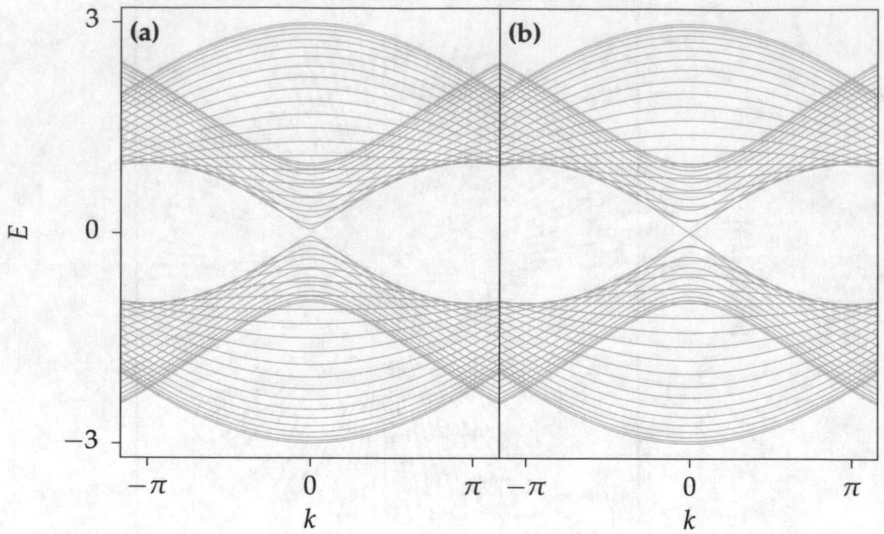

Fig. 2.17 **Subbands in armchair graphene nanoribbons. a** Semiconducting nanoribbon of $N_y = 30$ and **b** metallic nanoribbon of $N_y = 29$

$$q = \frac{p\pi}{N_y + 1}, \qquad p = 1, \ldots, N_y. \tag{2.3.44}$$

Upon substitution of $\psi_A(n)$ and $\psi_B(n)$ into Eq. (2.3.41) we can find that the spectrum is given by

$$E(k) = \pm\sqrt{1 + 2\epsilon_q \cos\left(\frac{k}{2}\right) + \epsilon_q^2}\,, \tag{2.3.45}$$

where $\epsilon_q = 2\cos(q)$. This result is in agreement to that discussed in Ref. [77]. An example for $N_y = 30$ is shown in Fig. 2.17a.

At $k = 0$, which is where $\Delta_a(\boldsymbol{k})$ touched the origin, we can write

$$E(0) = \pm|1 + \epsilon_q|\,. \tag{2.3.46}$$

The system can become metallic if $E(0) = 0$, which occurs for specific values of N_y. Indeed, for that matter we need $q = 2\pi/3$. This can be achieved whenever $N_y = 3r - 1$ with $r = 1, 2, \ldots$ since p runs from 1 to N_y and $2r$ is within 1 and $3r - 1$. This is shown in Fig. 2.17b for $N_y = 29$. Moreover, the dispersion of these metallic states close to $k = 0$ is massless Dirac-like. Indeed, if we expand $E(k)$ around $k = 0$ for the metallic states ($\epsilon_q = -1$), we obtain $E(k) = \pm k/2$, a Dirac cone. If we recall that k is expressed in units of the lattice constant and that $E(k)$ is expressed in units of t, we can write $E(k) = \pm v_F k$, where $v_F = 3ta_{cc}/2$, with a_{cc} the distance between neighbouring carbon atoms.

The metallic states are not protected by topology since they do not occur from a bulk-boundary correspondence. Instead, they occur due to choosing the right values for N_y such that upon slicing the bulk dispersion due to quantization of the transverse momentum we pass through the Dirac cone. However, since this Thesis focuses on the reshaping of Dirac cones, we shall be interested in these states and not on the zero energy modes that appear from topology in the zigzag and bearded nanoribbons. In order to conclude this chapter, we will present the solution to the armchair nanoribbon within the low-energy description, since it will be of importance later on. The discussion for the zigzag nanoribbon can be found in a number of Refs. [78–80]. We shall see that the low-energy description does provide a topological origin to the appearance of metallic behaviour [81]. In order to follow Ref. [81], we will write the low-energy Hamiltonian of the previous section in the slightly modified basis $\{\psi_A^K(x), \psi_B^K(x), -\psi_B^{K'}(x), \psi_A^{K'}(x)\}$, which simply means to change τ_z to τ_0 in the valley part of the Hamiltonian. Also, we will allow for the presence of a potential $V(x)$, since it will be of interest in the following chapter. Therefore, the Hamiltonian is written as follows

$$\mathcal{H} = v_F \tau_0 \otimes (\boldsymbol{\sigma}_\perp \cdot \boldsymbol{k}) + \tau_0 \otimes \sigma_0 V(x) \,. \tag{2.3.47}$$

It must be carefully remembered that $\boldsymbol{k}$ has to be considered as the momentum operator, which can be directly substituted by the quantum number $\boldsymbol{k}$ in translationally invariant systems. Also, it must be remembered that this Hamiltonian was obtained by rotating $\pi/2$ with respect to the lattice oriented as in Fig. 2.15. Hence, we have to consider that when we refer to such a figure, the direction along which the lattice is infinite corresponds to the Y-direction in our model and the direction along which it is finite corresponds to the X-direction [see Fig. 2.8]. Hence, $\boldsymbol{K} = -\boldsymbol{K}' = (4\pi/3)(1, 0)$. The important point here is how to impose the boundary conditions. One would be inclined to set to zero the envelope functions $\psi_\alpha^{K(K')}(\boldsymbol{r})$ at the edges. However, one has to take into account that the total wavefunction is actually written as follows [19] (cf. the $\boldsymbol{k} \cdot \boldsymbol{p}$ analysis of the SSH model)

$$\Psi(\boldsymbol{r}) = \sum_{\alpha=A,B} \sum_{\boldsymbol{R}_\alpha} \phi_\alpha(\boldsymbol{r}) \varphi(\boldsymbol{r} - \boldsymbol{R}_\alpha) \,, \tag{2.3.48}$$

where $\varphi(\boldsymbol{r} - \boldsymbol{R}_\alpha)$ is an atomic p_z orbital centered at $\boldsymbol{R}_\alpha$ and $\phi_\alpha(\boldsymbol{r})$ is written in terms of the envelope functions [80]

$$\phi_\alpha(\boldsymbol{r}) = e^{\mathrm{i}\,\boldsymbol{K}\cdot\boldsymbol{r}} \psi_\alpha^K(\boldsymbol{r}) + e^{\mathrm{i}\,\boldsymbol{K}'\cdot\boldsymbol{r}} \psi_\alpha^{K'}(\boldsymbol{r}) \,. \tag{2.3.49}$$

Notice the difference with the SSH model, where we expanded around $Q = \pi$. Here, the $\boldsymbol{K}$ and $\boldsymbol{K}'$ points are degenerate and $\phi_\alpha(\boldsymbol{r})$ has to take into account contributions from both points. This is why we obtain a 4×4 low-energy Hamiltonian, despite of having two bands only. Therefore, the proper boundary conditions in armchair nanoribbons imply that $\phi_A(\boldsymbol{r})$ and $\phi_B(\boldsymbol{r})$ must be zero at the boundary. This is not so for zigzag nanoribbons, where only one of the two sublattices has to be nullified at the boundary. A subtle point is where to precisely nullify the amplitudes ϕ_A and ϕ_B. We will do exactly the same as with the tight-binding model, where we nullified on a fictitious row of atoms right next to the proper boundary. As we said earlier, the system will be finite along the X-direction and infinite along the Y-direction. That allows us to have a good quantum number k along the Y-direction and therefore the problem in that direction is trivially solved by a phase factor $\exp(\mathrm{i}\,ky)$. As a result, we will write ϕ_A and ϕ_B to depend only on x, always remembering that they actually carry such a phase factor. Let us then call the position of the fictitious rows of atoms x_b, with $b = 1, 2$. In the nanoribbon shown in Fig. 2.15, this would correspond to $x_1 = 0$ and $x_2 = (N_x + 1)/2$, in units of the bulk graphene lattice constant $a = 1$. Notice that to agree with the results obtained in the tight-binding calculations, one has to make the substitution $N_y \to N_x$ in the results obtained earlier. Hence, $\phi_\alpha(x_b) = 0$ implies that

$$\psi_\alpha^K(x_b) = -e^{-\mathrm{i}2Kx_b}\psi_\alpha^{K'}(x_b)\,, \tag{2.3.50}$$

where we have taken into account that $\boldsymbol{K} = -\boldsymbol{K}'$. With this in mind, let us solve the problem in the absence of a potential. In the basis we have chosen, the Hamiltonian acts upon $\boldsymbol{\Psi} = (\psi_A^K, \psi_B^K, -\psi_B^{K'}, \psi_A^{K'})$, $\mathcal{H}\boldsymbol{\Psi} = E\boldsymbol{\Psi}$. Let us consider $V(x) = 0$. Then, if we square the Hamiltonian, we obtain

$$v_F^2(-\partial_x^2 + k^2)\boldsymbol{\Psi}(x) = E^2\boldsymbol{\Psi}(x)\,. \tag{2.3.51}$$

Therefore, each of the four components of $\boldsymbol{\Psi}(x)$ satisfies

$$v_F^2(-\partial_x^2 + k^2)\psi_\alpha^{K(K')}(x) = E^2\psi_\alpha^{K(K')}(x)\,. \tag{2.3.52}$$

However, not all four can be independent, since that would lead to eight integration constants instead of four. Indeed, the components within a valley subspace are actually coupled by $\boldsymbol{\sigma}_\perp \cdot \boldsymbol{k}$. Hence, we can obtain ψ_A^K and $\psi_A^{K'}$ from (2.3.52) and the other two would be obtained from

$$\psi_B^{K(K')}(x) = \mathrm{i}\,(\mp\partial_x + k)\,\psi_A^{K(K')}(x)\,, \tag{2.3.53}$$

where the minus sign corresponds to K and the plus sign to K'. We can propose the following ansatz

$$\psi_A^{K(K')}(x) = A_{K(K')}e^{\mathrm{i}qx} + B_{K(K')}e^{-\mathrm{i}qx}\,, \tag{2.3.54}$$

which implies that

$$\psi_B^{K(K')}(x) = (\mathrm{i}\,k \pm q)\,A_{K(K')}e^{\mathrm{i}qx} + (\mathrm{i}\,k \mp q)\,B_{K(K')}e^{-\mathrm{i}qx}\,. \tag{2.3.55}$$

Upon substitution of the ansatz into Eq. (2.3.52) leads to the following dispersion

$$E = \pm v_F\sqrt{k^2 + q^2}\,. \tag{2.3.56}$$

The allowed values for q are imposed by the boundary conditions. The boundary condition for ψ_A at $x_1 = 0$ implies that

$$A_K + B_K = -A_{K'} - B_{K'} . \qquad (2.3.57)$$

On the other hand, the boundary condition at $x_2 = (N_x + 1)/2$ implies

$$A_K e^{\mathrm{i}(q+K)x_2} + B_K e^{-\mathrm{i}(q-K)x_2} = -A_{K'} e^{\mathrm{i}(q-K)x_2} - B_{K'} e^{-\mathrm{i}(q+K)x_2} . \qquad (2.3.58)$$

These two equations have to be complemented with those for ψ_B, leading to a homogeneous system of equations with four unknowns. Upon imposing the determinant to be zero, we can obtain the allowed values for q. One has to be careful with the case $q = 0$, which has to be treated separately. Indeed, if $q = 0$, the term coming from the derivative in $\psi_B(x)$ is zero and, as a result, $\psi_B(x) = \mathrm{i}k\psi_A(x)$. Hence, two of those four equations would be redundant. In the case of having $q = 0$, it is straightforward to obtain from the boundary conditions that one must have

$$\exp(-\mathrm{i}\,2Kx_2) = 1 . \qquad (2.3.59)$$

Taking into account the value of K and x_2, this is equivalent to

$$\frac{2(N_x + 1)}{3} = n , \qquad n \in \mathbb{Z} , \qquad (2.3.60)$$

which can only occur if $N_x + 1$ is an integer multiple of 3. That is, $N_x = 3r - 1$ with $r = 1, 2, \ldots$. This result is in accordance to our results using the tight-binding formalism. On the other hand, setting the determinant to zero, excluding the $q = 0$ solution due to the previous reasoning, leads to the following equation

$$\sin[(q - K)x_2] \sin[(q + K)x_2] = 0 . \qquad (2.3.61)$$

Since the equation is symmetric when $q \to -q$, we can consider either one of the two terms and set it to zero. This way, we obtain that q is quantized

$$q_n = \frac{n\pi}{x_2} - K , \qquad n \in \mathbb{Z} . \qquad (2.3.62)$$

Notice that the case $q_n = 0$ is contained in this set of solutions when $N_x = 3r - 1$ with $r = 1, 2, \ldots$. We have therefore found that the dispersion given in Eq. (2.3.56) forms a set of subbands

$$E = \pm v_F \sqrt{k^2 + q_n^2} , \qquad q_n = \frac{n\pi}{x_2} - K , \qquad n \in \mathbb{Z} . \qquad (2.3.63)$$

In general, $q_n \neq 0$ and E forms a series of massive Dirac-like subbands. Since q_n is squared in the dispersion relation, one can wonder if there is any possibility of double degeneracy, that is, if there exists another integer n' such that $q_n = -q_{n'}$. If we take into account the values that q_n can take, this implies that

$$n + n' = \frac{4}{3}(N_x + 1) , \tag{2.3.64}$$

which is only satisfied if $N_x = 3r - 1$ with $r = 1, 2, \ldots$, which is the condition for having metallic behaviour. Hence, metallic nanoribbons are such that they have two singly degenerate massless Dirac bands corresponding to $q = 0, E = \pm k$, and a set of massive Dirac bands which are doubly degenerate. Semiconducting nanoribbons on the other hand have a set of massive Dirac bands only and these are non-degenerate. It is important to notice that this is only an effect that appears in the continuum description of graphene, and will be key to explain the topological protection of the Dirac cones in metallic nanoribbons when dealt with the Dirac equation. In the tight-binding description, this degeneracy becomes increasingly more precise the larger N_x is and the closer to $k = 0$ we are. This makes sense, since small values of N_x lead to large quantized transverse momenta, so the effects of *trigonal warping* cannot be neglected. These correspond to terms of order $\boldsymbol{k}^2$ in the bulk dispersion, which are accompanied by $\sin(3\theta_{\boldsymbol{k}})$, being $\theta_{\boldsymbol{k}} = \arctan(k_x/k_y)$ [37], which in turn lead to the dispersion not being isotropic around the Dirac points. It has threefold rotational symmetry, hence the adjective trigonal, but it breaks the symmetry between the K and K' points. We will see the relevance of this discussion in the next chapter.

As of now, we can try to understand the difference between metallic and semiconducting nanoribbons in the continuum description. One of such distinctions is that semiconducting nanoribbons display nondegenerate massive Dirac bands, whereas metallic nanoribbons have doubly degenerate massive Dirac bands and singly degenerate Dirac cones. We will show below that the degeneracy can be resolved due to the presence of a pseudovalley degree of freedom [81]. In order to reach these conclusions, we must rewrite the boundary conditions in the following form

$$\boldsymbol{\Psi}(x_b) = \mathcal{M}_b \boldsymbol{\Psi}(x_b) , \tag{2.3.65}$$

where

$$\mathcal{M}_b = \begin{pmatrix} 0 & -\mathrm{i}\, e^{\mathrm{i}\theta_b} \\ \mathrm{i}\, e^{-\mathrm{i}\theta_b} & 0 \end{pmatrix} \otimes \sigma_y , \tag{2.3.66}$$

and $\theta_b = -2Kx_b$. This matrix can also be written as follows

$$\mathcal{M}_b = \boldsymbol{u}_b \cdot \boldsymbol{\tau} \otimes \sigma_y , \qquad \boldsymbol{u}_b = (\sin\theta_b, \cos\theta_b, 0) . \tag{2.3.67}$$

In our configuration, $\theta_1 = 0$ and $\theta_2 = -4\pi(N_x + 1)/3$. Metallic nanoribbons have $N_x + 1 = 3r$ with $r \in \mathbb{Z}^+$, so that θ_1 and θ_2 are separated by an integer multiple of 4π. Therefore, $\mathcal{M}_1 = \mathcal{M}_2 \equiv \mathcal{M} = \tau_y \otimes \sigma_y$ in metallic nanoribbons. In this case, therefore, $\tau_y \otimes \sigma_0$ commutes with the low-energy Hamiltonian including the potential term, but also with $\mathcal{M}$. Therefore, we may choose solutions to be eigenstates of $\tau_y \otimes \sigma_0$, which in turn allows us to group them into two groups. Naively, one may think that the degeneracy comes from the Kramers' degeneracy between K and K' that we find in the bulk Hamiltonian. Indeed, in the bulk the Dirac cones that are obtained from the Hamiltonian, $E(\boldsymbol{k}) = \pm v_F |\boldsymbol{k}|$ are doubly degenerate since the Hamiltonian is a 4×4 matrix. However, Kramers' degeneracy comes from having an antiunitary symmetry that squares to -1 and one has to take great care when carrying the symmetries from the bulk into the finite systems. Indeed, for these symmetries to continue holding in the finite system, they must respect the boundary conditions as well. This amounts to having the boundary matrices $\mathcal{M}_b$ to commute with such symmetries [81]. We know of one of these antiunitary symmetries from our obtaining of the low-energy Hamiltonian, which is time-reversal symmetry. We wrote it as $\Theta = (\tau_x \otimes \sigma_x)\mathcal{K}$, with $\mathcal{K}$ being complex conjugation. In the present basis, we have to use rather $\Theta_y = (\tau_y \otimes \sigma_y)\mathcal{K}$. However, this symmetry squares to $+1$, since we chose it to be that for spinless fermions. Since the Hamiltonian is now diagonal in the valley subspace, we can propose three more antiunitary symmetries by taking into account that $(U \otimes \sigma_y)\mathcal{K}$ with U a unitary operator will also commute with the Hamiltonian. Hence, choosing U to be the remaining Pauli matrices, we have the following set of antiunitary symmetries that commute with $\mathcal{H}$ upon changing $\boldsymbol{k} \to -\boldsymbol{k}$

$$\Theta_i = \left(\tau_i \otimes \sigma_y\right) \mathcal{K} \,, \qquad i = 0, x, y, z \,. \tag{2.3.68}$$

That is,

$$\Theta_i \mathcal{H}(\boldsymbol{k}) \Theta_i^{-1} = \mathcal{H}(-\boldsymbol{k}) \,, \tag{2.3.69}$$

for all $i = 0, x, y, z$. Notice that the Θ_i's square to -1, with the exception of $i = y$ which squares to $+1$, as we already pointed out. It must be noted that Θ_0 and Θ_z are actually equivalent and belong to a larger class of intravalley operators differing each on a phase [81]

$$\Theta(\theta) = (\cos\theta\, \tau_0 + \mathrm{i}\, \sin\theta\, \tau_z) \otimes \sigma_y \mathcal{K} \,. \tag{2.3.70}$$

In the case of Θ_0 and Θ_z, these differ by a phase of $\pi/2$. Which one to choose from the family of operators will depend on the actual boundary conditions. It is the fact that Θ_x squares to -1 that leads to the Kramers' degeneracy of the two valleys in the bulk. It must be noted that the operators $\Theta(\theta)$ cannot be held responsible for such a degeneracy in the bulk, since they are intravalley operators and, therefore, do not mix the valley subspaces. With this in mind, we now have to check which of these symmetries commute as well with the boundary matrices $\mathcal{M}_1$ and $\mathcal{M}_2$. Since the symmetries have to commute with both boundary conditions, they have to commute in particular with $\mathcal{M}_1 = \tau_y \otimes \sigma_y$, which allows us to disregard by inspection Θ_x and Θ_z. In fact, it is easy to check that of the family $\Theta(\theta)$, the only one that commutes is $\Theta(0) = \Theta_0$. Θ_y also commutes with $\mathcal{M}_1$, so it remains to check if Θ_0 and Θ_y also commute with $\mathcal{M}_2$. If we do so, we find that Θ_y commutes with $\mathcal{M}_2$ for any value of θ_2, whereas Θ_0 only commutes with $\mathcal{M}_2$ if $u_2^x = 0$. This last condition can only be fulfilled if θ_2 is an integer multiple of 2π, which only occurs in metallic nanoribbons. We can then conclude the following: both metallic and semiconducting nanoribbons preserve the proper time-reversal symmetry of the bulk, but only the metallic nanoribbons carry another extra antiunitary symmetry from the bulk. However, this antiunitary symmetry is not Θ_x and, therefore, we cannot conclude the degeneracy of the subbands or the emergence of a Dirac cone in the metallic nanoribbons to be due to the Kramers' degeneracy of the valleys. However, if we check the product of these such symmetries, $\Theta_y \Theta_0 = -\tau_y \otimes \sigma_0$, we see that it is nothing but the matrix (up to an irrelevant minus sign) that we said to be commuting with both the Hamiltonian and the boundary matrix in the case of metallic nanoribbons. Hence, it is the fact that we have these two symmetries that allows us to group the solutions into two categories. These two categories can be labeled by a *pseudovalley index*. In order to obtain it, we can notice that while the valley states are eigenstates of $\tau_z \otimes \sigma_0$, the pseudovalley states are eigenstates of $\tau_y \otimes \sigma_0$, so the latter can be obtained from a linear combination of the former. Therefore, if we want to connect the known result from the bulk where K and K' are Kramers' partners, that is, $E_K(\boldsymbol{k}) = E_{K'}(-\boldsymbol{k})$, we just have to perform a $\pi/2$ rotation so that we carry τ_y to τ_z around the valley X-axis [81]. We do so by means of the following operator

$$\mathcal{R} = \exp\left(-\mathrm{i}\,\frac{\pi}{4}\tau_x\right) \otimes \sigma_0 = \frac{1}{\sqrt{2}}\left(\tau_0 - \mathrm{i}\,\tau_x\right) \otimes \sigma_0 \,. \tag{2.3.71}$$

Upon rotation, Θ_y remains unchanged, whereas Θ_0 changes to

$$\Theta_0^{\mathcal{R}} = -\mathrm{i}\,\tau_x \otimes \sigma_y\,\mathcal{K}\,. \tag{2.3.72}$$

On the other hand, the matrix $\tau_y \otimes \sigma_0$ changes to $\sigma_z \otimes \sigma_0$, as we desired to. Finally, the boundary matrix changes to

$$\mathcal{M}^{\mathcal{R}} = \tau_z \otimes \sigma_y\,. \tag{2.3.73}$$

Thus, if we denote the pseudovalley indices by $K_{\mathcal{R}}$ and $K'_{\mathcal{R}}$, we can conclude that the spectrum can be pseudovalley resolved

$$E_{K_{\mathcal{R}}}(k) = E_{K'_{\mathcal{R}}}(-k)\,. \tag{2.3.74}$$

This implies that, if they exist, gapless Dirac-like modes are Kramers' partners to one another and are singly degenerate, whereas massive subbands are doubly degenerate. A schematic depiction of the pseudovalley resolution is shown in Fig. 2.18.

The existence of gapless modes can be traced back to the form of $\mathcal{M}^{\mathcal{R}}$. For that matter, we have to discuss first what are called the *infinite-mass boundary conditions*. These were introduced by Berry and Mondragon [82] when trying to work out the problem of confinement of massless spin-1/2 particles. They showed that the way to confine these was to introduce a mass or scalar potential term which tends towards

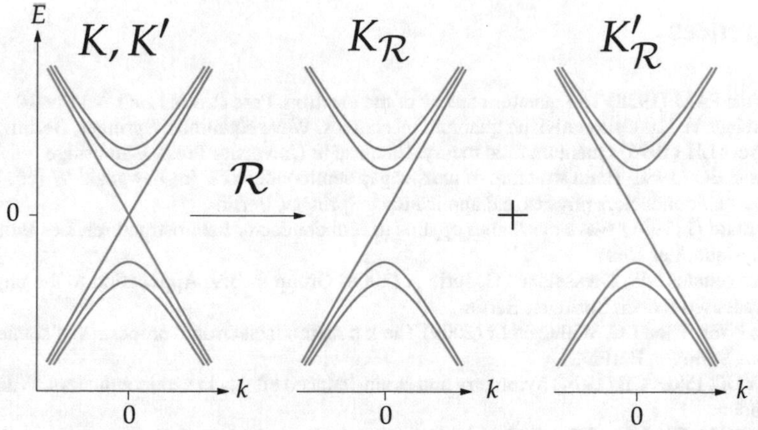

Fig. 2.18 **Pseudovalley resolved subbands in metallic armchair nanoribbons**. A rotation $\mathcal{R}$ in the valley subspace allows us to unravel a pseudovalley structure of the continuum model [81]

infinity outside the confinement region and is zero within it. Hence the name infinite-mass boundary conditions. In the end, they showed that by requiring the probability current to vanish outside the confinement region, one could write at the boundary [83]

$$\boldsymbol{\Psi} = -\mathcal{M}_\infty^b \boldsymbol{\Psi} , \qquad \mathcal{M}_\infty^b = \tau_z \otimes \left(\boldsymbol{\sigma} \cdot \left[\boldsymbol{n}_b \times \hat{z}\right]\right) , \tag{2.3.75}$$

where $\boldsymbol{n}_b$ is a vector normal to the boundary. In our case, $\boldsymbol{n}_1 = -\boldsymbol{n}_2 = -\hat{x}$ which would imply

$$\mathcal{M}_\infty^1 = -\mathcal{M}_\infty^2 = \tau_z \otimes \sigma_y . \tag{2.3.76}$$

This has actually the same form as $\mathcal{M}^{\mathcal{R}}$, but the crucial difference is that $\mathcal{M}^{\mathcal{R}}$ is the same on both sides, whereas the infinite-mass condition has $\mathcal{M}_\infty$ of opposite sign on opposite boundaries. We may then conclude that the boundary conditions of metallic armchair nanoribbons are infinite-mass conditions, with *opposite masses on each side* [81]. This should sound familiar, since we have already encountered it twice, once in the SSH and another at the topological boundary: a change in the sign of the mass creates a domain wall and leads to gapless modes. In one dimension, these were zero modes, in three dimensions these where two-dimensional Dirac cones. Therefore, we may argue that the origin for the Dirac cones in metallic nanoribbons is actually topological. What is more interesting is that this analysis holds even in the presence of a potential $V(\boldsymbol{r})$, meaning that the Dirac point will be protected even in presence of a nonzero $V(\boldsymbol{r})$. We shall see the importance of this result in the following chapter.

References

1. Dirac PAM (1928) The quantum theory of the electron. Proc R Soc Lond A 117:610
2. Greiner W (2000) Relativistic quantum mechanics. Wave equations. Springer, Berlin
3. Ryder LH (1996) Quantum field theory. Cambridge University Press, Cambridge
4. Kane EO (1980) Band structure of narrow gap semiconductors. In: Zawadzki W (ed) Narrow gap semiconductors physics and applications. Springer, Berlin
5. Bastard G (1991) Wave mechanics applied to semiconductor heterostructures. Les Editions de Physique, Les Ulis
6. Dresselhaus MS, Dresselhaus G, Jorio A (2008) Group theory. Application to the physics of condensed matter. Springer, Berlin
7. Lew Yan Voon LC, Willatzen M (2009) The k p method: electronic properties of semiconductors. Springer, Berlin
8. Bir GL, Pikus GE (1974) Symmetry and strain-induced effects in semiconductors. Wiley, New York
9. Ivchenko EL, Pikus GE (1997) Superlattices and other heterostructures: symmetry and optical phenomena. Springer, Berlin
10. Winkler R (2003) Spin-orbit coupling effects in two-dimensional electron and hole systems. Springer, Berlin
11. Nakahara M (2003) Geometry, topology and physics. Taylor & Francis, Boca Raton
12. Aharonov Y, Bohm D (1959) Significance of electromagnetic potentials in the quantum theory. Phys Rev 115:485
13. Su WP, Schrieffer JR, Heeger AJ (1979) Solitons in polyacetylene. Phys Rev Lett 42:1698
14. Peierls RE (1955) Quantum theory of solids. Oxford University Press, Oxford
15. Peierls RE (1991) More surprises in theoretical physics. Princeton University Press, Oxford
16. Asbóth JK, Oroszlány L, Pályi A (2016) A short course on topological insulators: band structure and edge states in one and two dimensions. Springer International Publishing, Heidelberg
17. Ballentine LE (2000) Quantum mechanics: a modern development. World Scientific Publishing, Singapore
18. Franz M, Molenkamp L (2013) Topological insulators. Elsevier, Oxford
19. Marconcini P, Macucci M (2011) The k p method and its application to graphene, carbon nanotubes and graphene nanoribbons: the Dirac equation. Riv Nuovo Cimento 34:489
20. Zhang F, Kane CL, Mele EJ (2012) Surface states of topological insulators. Phys Rev B 86:081303
21. Haldane FDM (1988) Model for a quantum Hall effect without Landau levels: condensed-matter realization of the parity anomaly. Phys Rev Lett 61:2015
22. Kane CL, Mele EJ (2005) Z_2 topological order and the quantum spin Hall effect. Phys Rev Lett 95:146802
23. Hatsugai Y (1993) Chern number and edge states in the integer quantum Hall effect. Phys Rev Lett 71:3697
24. Wen X-G, Wub Y-S, Hatsugai Y (1994) Chiral operator product algebra and edge excitations of a fractional quantum Hall droplet. Nucl Phys B 422:476
25. Bernevig AB, Hughes T (2013) Topological insulators and topological superconductors. Princeton University Press, New Jersey
26. Jackiw R, Rebbi C (1976) Solitons with fermion number 1/2. Phys Rev D 13:3398
27. Winkler R, Zülicke U (2010) Invariant expansion for the trigonal band structure of graphene. Phys Rev B 82:245313
28. Cornwell JF (1997) Group theory in physics. Academic, San Diego
29. Chiu C-K, Teo JCY, Schnyder AP, Ryu S (2016) Classification of topological quantum matter with symmetries. Rev Mod Phys 88:035005
30. Altland A, Zirnbauer MR (1997) Nonstandard symmetry classes in mesoscopic normal-superconducting hybrid structures. Phys Rev B 55:1142
31. Bernard D, LeClair A (2002) A classification of 2D random Dirac fermions. J Phys A: Math Gen 35:2555

32. Kitaev A (2009) Periodic table for topological insulators and superconductors. AIP Conf Proc 1134:22
33. Schnyder AP, Ryu S, Furusaki A, Ludwig AWW (2008) Classification of topological insulators and superconductors in three spatial dimensions. Phys Rev B 78:195125
34. Pérez-González B, Bello M, Gómez-León Á, Platero G (2019) Interplay between long-range hopping and disorder in topological systems. Phys Rev B 99:035146
35. Rice MJ, Mele EJ (1982) Elementary excitations of a linearly conjugated diatomic polymer. Phys Rev Lett 49:1455
36. Shen S-Q (2012) Topological insulators: Dirac equation in condensed matters. Springer, Berlin
37. Castro Neto AH, Guinea F, Peres NMR, Novoselov KS, Geim AK (2009) The electronic properties of graphene. Rev Mod Phys 81:109
38. Von Neumann J, Wigner EP (1929) Über merkwürdige diskrete Eigenwerte. Phys Z 30:467
39. Berry MV (1984) Quantal phase factors accompanying adiabatic changes. Proc R Soc Lond A 392:45
40. Hatsugai Y, Fukui T, Aoki H (2007) Topological aspects of graphene. Eur Phys J Spec Top 148:133
41. Montambaux G, Piéchon F, Fuchs J-N, Goerbig MO (2009) Merging of Dirac points in a two-dimensional crystal. Phys Rev B 80:153412
42. Hatsugai Y (2011) Topological aspect of graphene physics. J Phys: Conf Ser 334:012004
43. Semenoff GW (1984) Condensed-matter simulation of a three-dimensional anomaly. Phys Rev Lett 53:2449
44. Kane CL, Mele EJ (2005) Quantum spin Hall effect in graphene. Phys Rev Lett 95:226801
45. Hasan MZ, Kane CL (2010) Colloquium: topological insulators. Rev Mod Phys 82:3045
46. König M, Buhmann H, Molenkamp LW, Hughes T, Liu C-X, Qi X-L, Zhang S-C (2008) The quantum spin Hall effect: theory and experiment. J Phys Soc Jpn 77:031007
47. Qi X-L, Zhang S-C (2011) Topological insulators and superconductors. Rev Mod Phys 83:1057
48. Yan B, Zhang S-C (2012) Topological materials. Rep Prog Phys 75:096501
49. Ando Y (2013) Topological insulator materials. J Phys Soc Jpn 82:102001
50. Wen X-G (2017) Colloquium: zoo of quantum-topological phases of matter. Rev Mod Phys 89:041004
51. DiVincenzo DP, Mele EJ (1984) Self-consistent effective-mass theory for intralayer screening in graphite intercalation compounds. Phys Rev B 29:1685
52. Kechedzhi K, McCann E, Fal'ko VI, Suzuura H, Ando T, Altshuler BL (2007) Weak localization in monolayer and bilayer graphene. Eur Phys J Spec Top 148:39
53. Zhang H, Liu C-X, Qi X-L, Dai X, Fang Z, Zhang S-C (2009) Topological insulators in Bi_2Se_3, Bi_2Te_3 and Sb_2Te_3 with a single Dirac cone on the surface. Nat Phys 5:438
54. Fu L, Kane CL (2007) Topological insulators with inversion symmetry. Phys Rev B 76:045302
55. Liu C-X, Qi X-L, Zhang H, Dai X, Fang Z, Zhang S-C (2010) Model Hamiltonian for topological insulators. Phys Rev B 82:045122
56. Ortmann F, Roche S, Valenzuela SO (2015) Topological insulators. Wiley, New York
57. Rössler U (2009) Solid state theory: an introduction. Springer, Berlin
58. Dimmock JO, Wright GB (1964) Band edge structure of PbS, PbSe, and PbTe. Phys Rev 135:A821
59. Dimmock JO, Melngailis I, Strauss AJ (1966) Band structure and laser action in $Pb_xSn_1 - x$Te. Phys Rev Lett 16:1193
60. Mitchell DL, Wallis RF (1966) Theoretical energy-band parameters for the lead salts. Phys Rev 151:581
61. Nimtz G, Schlicht B (1983) Narrow-gap semiconductors. Springer, Berlin
62. Volkov BA, Pankratov OA (1985) Two-dimensional massless electrons in an inverted contact. Sov Phys JETP 42:178
63. Pankratov OA, Pakhomov SV, Volkov BA (1987) Supersymmetry in heterojunctions: band-inverting contact on the basis of $Pb_{1-x}Sn_x$Te and $Hg_{1-x}Cd_x$Te. Solid State Commun 61:93
64. Korenman V, Drew HD (1987) Subbands in the gap in inverted-band semiconductor quantum wells. Phys Rev B 35:6446

65. Agassi D, Korenman V (1988) Interface states in band-inverted semiconductor heterojunctions. Phys Rev B 37:10095
66. Pankratov OA (1990) Electronic properties of band-inverted heterojunctions: supersymmetry in narrow-gap semiconductors. Semicond Sci Technol 5:S204
67. Domínguez-Adame F (1994) Green function approach to interface states in band-inverted junctions. Phys Status Solidi (b) 186:K49
68. Fu L (2011) Topological crystalline insulators. Phys Rev Lett 106:106802
69. Hsieh TH, Lin H, Liu J, Duan W, Bansil A, Fu L (2012) Topological crystalline insulators in the SnTe material class. Nat Commun 3:982
70. Teo JCY, Fu L, Kane CL (2008) Surface states and topological invariants in three-dimensional topological insulators: application to $Bi_{1-x}Sb_x$. Phys Rev B 78:045426
71. Thouless DJ, Kohmoto M, Nightingale MP, den Nijs M (1982) Quantized Hall conductance in a two-dimensional periodic potential. Phys Rev Lett 49:405
72. Ando Y, Fu L (2015) Topological crystalline insulators and topological superconductors: from concepts to materials. Annu Rev Condens Matter Phys 6:361
73. Tchoumakov S, Jouffrey V, Inhofer A, Plaçais B, Carpentier D, Goerbig MO (2017) Volkov-Pankratov states in topological heterojunctions. Phys Rev B 96:201302
74. Inhofer A, Tchoumakov S, Assaf BA, F'eve G, Berroir JM, Jouffrey V, Carpentier D, Goerbig MO, Plaçais B, Bendias K, Mahler DM, Bocquillon E, Schlereth R, Brüne C, Buhmann H, Molenkamp LW (2017) Observation of Volkov-Pankratov states in topological HgTe heterojunctions using high-frequency compressibility. Phys Rev B 96:195104
75. Hatsugai Y (1997) Topological aspects of the quantum Hall effect. J Phys: Condens Matter 9:2507
76. Ryu S, Hatsugai Y (2002) Topological origin of zero-energy edge states in particle-hole symmetric systems. Phys Rev Lett 89:077002
77. Wakabayashi K, Sasaki K, Nakanishi T, Enoki T (2010) Electronic states of graphene nanoribbons and analytical solutions. Sci Technol Adv Mater 11:054504
78. Brey L, Fertig HA (2006) Electronic states of graphene nanoribbons studied with the Dirac equation. Phys Rev B 73:235411
79. Wakabayashi K, Takane Y, Yamamoto M, Sigrist M (2009) Electronic transport properties of graphene nanoribbons. New J Phys 11:095016
80. Wurm J, Wimmer M, Adagideli İ, Richter K, Baranger HU (2009) Interfaces within graphene nanoribbons. New J Phys 11:095022
81. Wurm J, Wimmer M, Richter K (2012) Symmetries and the conductance of graphene nanoribbons with long-range disorder. Phys Rev B 85:245418
82. Berry MV, Mondragon RJ (1987) Neutrino billiards: time-reversal symmetry-breaking without magnetic fields. Proc R Soc Lond A Math Phys Sci 412:53
83. Akhmerov AR, Beenakker CWJ (2008) Boundary conditions for Dirac fermions on a terminated honeycomb lattice. Phys Rev B 77:085423

Chapter 3
Reshaping of Dirac Cones by Electric Fields

In the previous chapter, we discussed how topology can be interwoven within the physics of certain materials and how it produces nonstandard behaviours regarding the dispersion of electrons in such materials. Indeed, we saw that by placing two insulators of opposite $\mathbb{Z}_2$ invariant, creating a topological boundary, leads to the presence of topological surface states at the boundary. The dispersion of such states is a Dirac cone and the Dirac point is protected by time-reversal symmetry in three-dimensional topological insulators. On the other hand, we discussed that armchair graphene nanoribbons of certain width display also Dirac-like states at low energy and we could observe that, in the continuum description, these states are topologically robust, even in the presence of a spatially dependent potential. The protection stems from the presence of antiunitary symmetries that produce a pseudovalley structure, the two pseudovalleys being Kramers' partners.

In this chapter, we will consider those two systems when a uniform electric field is applied perpendicularly to the conserved momenta. That is, the electric field will be perpendicular to the surface of the topological insulator and it will be contained within the nanoribbon and pointing along its finite direction. We shall observe that such a perturbation, which does not break time-reversal symmetry, nor does it break the antiunitary symmetries of the nanoribbon, preserves the Dirac point, while widening the cone, thereby modifying the Fermi velocity [1]. The starting point will be to consider the effect of the electric field at a topological boundary by means of first order perturbation theory. Next, we will focus on the centered-symmetric boundary and will solve the problem under some approximations to the lowest non-zero order in the electric field. Within these approximations, it also makes sense to tackle the analysis of the effect of the electric field upon two topological boundaries, separated a certain distance. In this case, the net effect is the same as with the H_2^+ molecule, where two localized orbitals are brought together and, upon overlapping, lead to bonding and antibonding orbitals. Therefore, the Dirac dispersion will be gapped out and the gap will be tunable by the electric field [2]. These two approximate methods will shed light on the full problem and will allow us to obtain analytic expressions of the dispersion relation for low enough fields. To conclude the part

Á. Díaz Fernández, *Reshaping of Dirac Cones in Topological Insulators and Graphene*, Springer Theses, https://doi.org/10.1007/978-3-030-61555-0_3

devoted to the topological boundary, we will present an exact solution for arbitrary sized boundaries, which does not lead to analytic results, although it predicts the exact same phenomenon of the approximate calculations in the limit of low fields [3]. After this thorough study, we will drive our attention to narrow metallic armchair nanoribbons. As we know, the topological protection within the continuum model will protect the Dirac point. The result of the electric field will be to widen the cone and therefore alter the Fermi velocity, as with the other Dirac material studied. We shall see that the same widening happens in the tight-binding description, although the absence of topological protection here leads to a dispersion that is gapped out. However, the fact that the ribbons are narrow implies that the subbands are well separated from each other and we can go higher up in energy where the spectrum is again linear, and compare with the results of the continuum model. The agreement is noteworthy, as we shall observe.

3.1 Topological Boundary

3.1.1 Perturbation Theory

In the previous chapter, we saw that three dimensional topological insulators such as Bi_2Se_3 and topological crystalline insulators such as SnTe, can be described by a low-energy Hamiltonian that resembles the Dirac equation with a mass term given by half the energy gap. We then saw that by placing such materials next to a trivial insulator leads to the appearance of gapless, Dirac-like states at the surface. As we saw in that chapter, after introducing several dimensionless variables, the Hamiltonian for the topological boundary can be written as follows

$$\mathcal{H}_0 = -\mathrm{i}\,\alpha_z \partial_\xi + \boldsymbol{\alpha}_\perp \cdot \boldsymbol{\kappa} + \delta\beta + (\beta + \gamma)\,\mathrm{sgn}\,(\xi). \tag{3.1.1}$$

For a reminder on the definition of the dimensionless variables and their relation to the real material parameters, the reader is referred to the previous chapter, in particular to the section devoted to the topological boundary. As a brief reminder, δ and γ measure how different in size are the energy gaps and what is the separation between their centres, with $\delta = \gamma = 0$ corresponding to the same-sized, symmetric system. ξ is the direction perpendicular to the surface and it is measured in units of $d = v_z/\lambda$, being λ the average gap. The energies, ε, are measured in units of λ. $\boldsymbol{\kappa}$ is the in-plane momentum measured in units of $1/d$. Finally, the Hamiltonian is written in the orbital-spin basis and it acts upon the bispinor $\boldsymbol{\Psi}^0(\xi)$ containing the envelope functions

$$\mathcal{H}_0 \boldsymbol{\Psi}^0(\xi) = \left(\varepsilon^0 - \gamma\delta\right) \boldsymbol{\Psi}^0(\xi). \tag{3.1.2}$$

The superscript 0 indicates the unperturbed system. It is important to remember that the actual bispinor carries a phase $\exp(\mathrm{i}\,\boldsymbol{\kappa} \cdot \boldsymbol{r}_\perp)$ due to translational symmetry in the

XY-plane (recall that $\boldsymbol{\kappa}$ is the in-plane momentum, so its Z-component is nullified). This system hosts topological surface states that are localized at the surface

$$\boldsymbol{\Psi}^0_{\kappa,s}(\xi) = \exp\left[-\pi_0(\xi)|\xi|\right] \boldsymbol{\Phi}^0_{\kappa,s}, \tag{3.1.3}$$

where $s = \pm$ and

$$\pi_0^2(\xi) = \kappa^2 + (\delta + \mathrm{sgn}\,(\xi))^2 - \left(\varepsilon^0_{\kappa,s} - \gamma \mathrm{sgn}\,(\xi) - \gamma\delta\right)^2. \tag{3.1.4}$$

Notice that the decay lengths on both sides in units of d are given by

$$\ell^R_{\kappa,s} = \frac{1}{\pi_0^+(\kappa, s)}, \qquad \ell^L_{\kappa,s} = \frac{1}{\pi_0^-(\kappa, s)}, \tag{3.1.5}$$

where $\pi_0^\pm = \pi_0(0^\pm)$ and we have noted that, in general, $\pi_0^\pm$ depends on κ and s. The dispersion of the surface states is a single Dirac cone

$$\varepsilon^0_{\kappa,s} = s\sqrt{1-\gamma^2}\,\kappa. \tag{3.1.6}$$

Remember that $\kappa = |\boldsymbol{\kappa}|$. The double degeneracy at $\boldsymbol{\kappa} = 0$ is due to Kramers' theorem, since that is the only time-reversal-symmetric momentum of this model. Away from $\boldsymbol{\kappa} = 0$, Kramers' theorem does not hold anymore, the degeneracy splits and leads to the Dirac cone. In fact, the constant vector $\boldsymbol{\Phi}^0_{\kappa,s}$ is obtained as a linear superposition of the time-reversed partners at $\boldsymbol{\kappa} = 0$.

In order to understand future results in the case of an applied electric field, let us discuss Eq. (3.1.6) and relate it with the decay lengths on both sides. The first thing we notice is that for centered gaps, $\gamma = 0$, the decay lengths are independent of κ. That is, all states decay in exactly the same manner, independent of κ, even the Dirac point. They do not, however, decay equally fast on both sides, that depends on δ. Indeed, if $\delta \in (0, 1)$, it will decay faster to the right, in order to guarantee the normalization of the state, and the opposite will happen if $\delta \in (-1, 0)$. This makes sense if we consider that the gap is given by

$$\varepsilon_G(\xi) = 2\left[\delta + \mathrm{sgn}\,(\xi)\right]. \tag{3.1.7}$$

Hence, if $\delta > 0$, $|\varepsilon_G(0^+)| > |\varepsilon_G(0^-)|$. The larger the energy gap is, the smaller the decay length, and so $\ell_R < \ell_L$. The situation described above is schematically shown in Fig. 3.1a. The reason for the decay lengths not to be dependent on κ can be understood by considering the bulk and surface dispersions. Indeed, if $\gamma = 0$, the bulk dispersion on each side would be

$$\varepsilon^b_j = s\sqrt{\kappa^2 + \left(\delta + r_j\right)^2}, \qquad j = \mathrm{L, R}, \tag{3.1.8}$$

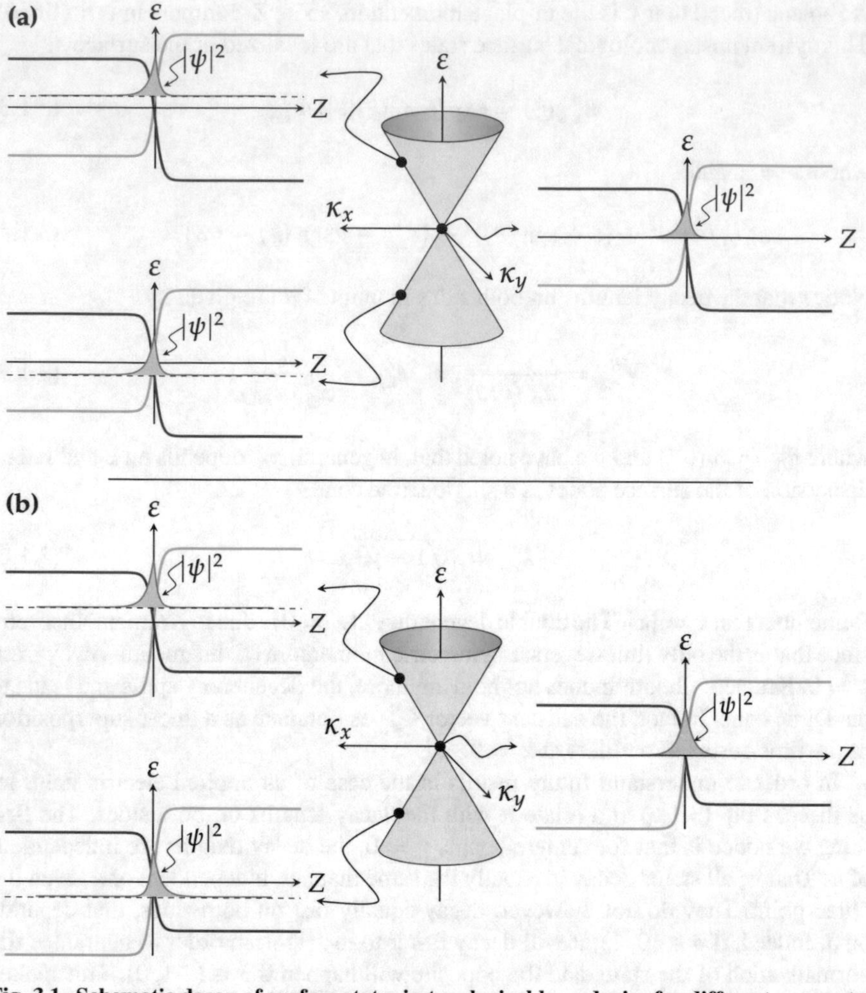

Fig. 3.1 Schematic decay of surface states in topological boundaries for different momenta. **a** Centered, asymmetric scenario. Surface states decay faster on the largest-gap side, independent of momentum. **b** Symmetric, off-centered situation. Surface states decay faster (slower) to the right the higher (lower) we go in energy with respect to the decay to the left. In both **a** and **b**, above and below the band edges, there is a continuum of states

with $r_\mathrm{L} = -1$ and $r_\mathrm{R} = 1$. It is clear that the surface state dispersion, $\varepsilon^0_{\kappa,s} = s\kappa$, would never cross the bulk dispersion, since at all times in the range $\delta \in (0, 1)$ we have $\varepsilon^b_j > \varepsilon^0_{\kappa,s}$.

Let us now consider the opposite situation, that with $\delta = 0$ and $\gamma \neq 0$. In this case, we can observe that the decay lengths do depend on κ. Again, we can look at the bulk dispersion, which in this case is given by

$$\varepsilon_j^b = r_j\gamma + s\sqrt{\kappa^2+1}. \tag{3.1.9}$$

As we can observe, if $\gamma > 0$, the term $r_j\gamma$ is positive on the right and negative on the left. Since the squared-root term tends asymptotically to κ, then there will be a point where $\varepsilon_{\mathrm{L}}^b < \kappa$, meaning that the dispersions would cross. In order to avoid such a situation, the less restrictive requirement is to ask for the cone to widen so that $\varepsilon_{\mathrm{L}}^b$ only touches the surface dispersion tangentially. For that matter, we write the surface dispersion as $\alpha\kappa$ and find α. If we do so, we find that $\alpha = \sqrt{1-\gamma^2}$. This is precisely what we found earlier, see Eq. (3.1.6). This situation is achieved by having κ-dependent decay lengths. If we consider $\gamma \in (0, 1)$, positive energy states of larger κ would be closer than states of lower κ to the crossing point with the bulk states on the left. Therefore, the decay lengths of the former become larger on that side. The opposite would happen for states of negative energies. In fact, it is only the Dirac point states that decay exactly the same on both sides because its proximity to the crossing points is the same with respect to both sides. However, if we consider the range where $\gamma \in (0, 1)$, states of larger momenta that are higher (lower) in energy with respect to the Dirac point decay faster (slower) to the right than to the left. This is shown schematically in Fig. 3.1b. In order to avoid the crossing with bulk states, higher (lower) energy states will move towards lower (higher) energies. Notice that states of larger momenta will suffer this fate more than states of lower momenta, meaning that the displacement in energy depends on $\boldsymbol{\kappa}$. However, momentum has to be conserved and rotation symmetry still dictates that the energy has to depend only on κ. The net result is a dispersion which is that of a Dirac cone for centered gaps but widened, which in turn translates into having a Fermi velocity that is reduced by $\sqrt{1-\gamma^2}$. In fact, in the limiting case where $\gamma \to 1$, the dispersion flattens since the overlap between the gaps on both sides shrinks to zero. Notice that the same result occurs if $\gamma \in (-1, 0)$, which is why the reduction goes with γ^2.

After this discussion, let us introduce a uniform electric field along the Z-direction. That is, let us add a potential of the form $V(\xi) = f\xi$, where f is the electric field strength measured in units of λ/ed. This allows us to introduce an *electric length*

$$\ell_F = \frac{\lambda}{e|F|}, \tag{3.1.10}$$

where F is the electric field strength with dimensions. Notice that $f = \mathrm{sgn}\,(F)d/\ell_F$. This way, one can distinguish two regimes: low fields, where $\ell_F \gg d$, and high fields, where $\ell_F \ll d$. In fact, one can introduce a *critical electric length*, ℓ_F^C, such that the corresponding *critical field*, F_C, leads to $\ell_F^C = d$. That is, $F_C = \lambda/ed$. We shall focus on the first regime, where the Dirac state is ensured to survive to the perturbation. Indeed, ℓ_F corresponds to the length across which the potential drop is λ and, if this is smaller than the critical electric length, the Dirac point will be far from the bulk states and tunneling into the continuum will be exponentially suppressed. We shall also see in the next section that this is indeed the case and that, in fact, it corresponds to a sensible experimental regime.

With this in mind, the Hamiltonian is now given by

$$\mathcal{H} = \mathcal{H}_0 + f\xi. \tag{3.1.11}$$

The first thing one may ask when using first-order perturbation theory is whether we should use degenerate or non-degenerate perturbation theory. On the one hand, there is Kramers' degeneracy. Since the potential does not break time-reversal symmetry, this degeneracy is not affected by the potential. Moreover, the scattering element $\langle \Psi^0 | V | \mathcal{T}\Psi^0 \rangle = 0$, since $V(\xi)$ is proportional to the identity and the Kramers' partners, $\boldsymbol{\Psi}^0$ and $\mathcal{T}\boldsymbol{\Psi}^0$, are orthogonal to each other. On the other hand, there is the degeneracy that arises due to rotational symmetry, which is responsible for the energy depending exclusively on κ. That is, the symmetry that leads to having an isotropic cone. However, since the potential is dependent only on the coordinate perpendicular to the surface, states of different $\boldsymbol{\kappa}$ are not mixed by the potential. In other words, plane waves of different in-plane momenta are orthogonal to each other and are not mixed by a potential that only contains the coordinate perpendicular to such momenta. Therefore, we can make use of first-order, non-degenerate perturbation theory, in order to assess the effects of the potential $V(\xi)$. Hence,

$$\varepsilon_{\kappa,s} = \varepsilon^0_{\kappa,s} + f\langle \xi \rangle^0_{\kappa,s}. \tag{3.1.12}$$

Notice that $\varepsilon_{\kappa,s}$ still depends only on κ since rotational symmetry at the surface is preserved. Since $f = \text{sgn}\,(F)/\ell_F$ (assuming ℓ_F in units of d), we can write the equation above as

$$\varepsilon_{\kappa,s} = \varepsilon^0_{\kappa,s} + \text{sgn}\,(F\langle \xi \rangle^0_{\kappa,s})\frac{|\langle \xi \rangle^0_{\kappa,s}|}{\ell_F}. \tag{3.1.13}$$

This equation is particularly appealing. In order to see this, let us start by analyzing the $\text{sgn}\,(F\langle \xi \rangle^0_{\kappa,s})$. Assume that $F > 0$ and $\langle \xi \rangle^0_{\kappa,s} < 0$ for a given state with momentum κ in the branch s. In this case, $\text{sgn}\,(F\langle \xi \rangle^0_{\kappa,s}) < 0$ and the original energy level will move downwards in energy. Let us try to understand why this is so. If we imagine the band edges as tilting due to the electric potential, then the first assumption implies that the band edges tilt with a positive slope. As a result, the continuum states on the left of the interface will be lower in energy than those to the right of the interface. The second assumption implies that we expect to find an electron in the state $(\boldsymbol{\kappa}, s)$ to be on the lefthand side of the junction in the unbiased system. Hence, this corresponds to a situation where the state decays slower to the left than to the right. From our previous discussion of the unbiased system, this occurs whenever the state is closer to a band edge on the left. For instance, in the centered, asymmetric scenario of Fig. 3.1a it occurs both for positive and negative energies, whereas for the symmetric, off-centered case of Fig. 3.1b it occurs for positive energies only. In either case, upon band tilting, there will be continuum states at the same energy of the surface states that they can resonate with on both sides of the interface. However, those on the left will be closer and hybridization with these states becomes more likely. In order to overcome this effect and protect the surface state, it will displace downwards in energy, so that it is farther apart from the states in the continuum. How much will it displace depends on how asymmetric the probability density is, that is, on the size of $\langle\xi\rangle^0_{\kappa,s}$. This is because when an energy state moves in energy to compensate for the proximity of the state to the bulk levels it has to pay the price of moving the state closer to the bulk levels on the other side. Hence, if $|\langle\xi\rangle^0_{\kappa,s}|$ is not too large, meaning that it is approximately equally likely to find an electron in either side of the interface, the change in energy cannot be too large for it would penalize the side that was originally better off. In fact, the symmetric-centered scenario has no shift at all precisely due to this effect. The shift in energy will also depend on how much the band edges tilt. Indeed, a large tilt implies that the continuum states will be closer to the Dirac point, as measured by the electric length ℓ_F. Hence, the displacement will be inversely proportional to ℓ_F to first order.

All of these considerations explain the result of Eq. (3.1.13) in a heuristic manner. In order to make it more clear, we shall calculate $\langle\xi\rangle^0_{\kappa,s}$ and apply the results to the band edge configurations that we discussed in the unbiased system. If we write $\pi_0(\xi) = \pi_0^+\Theta(\xi) + \pi_0^-\Theta(-\xi)$, where $\pi_0^\pm = \pi_0(0^\pm)$ and $\Theta(\xi)$ is the Heaviside step function, then the expectation values above can be analytically obtained and the result is the following

$$\langle\xi\rangle^0_{\kappa,s} = \frac{\ell^R_{\kappa,s} - \ell^L_{\kappa,s}}{2}. \tag{3.1.14}$$

As we can see, the intuition behind $\langle\xi\rangle^0_{\kappa,s}$ is exactly the one given above and it is easily thought of as the average between the decay lengths (with sign). We can understand its magnitude and sign from a competition between $\ell^R_{\kappa,s}$ and $\ell^L_{\kappa,s}$. As a result, we find

$$\varepsilon_{\kappa,s} = \varepsilon^0_{\kappa,s} + \operatorname{sgn}(F)\frac{\ell^R_{\kappa,s} - \ell^L_{\kappa,s}}{2\ell_F}. \tag{3.1.15}$$

As we discussed previously, what is actually relevant in determining where will the level go to is the sign of $\operatorname{sgn}(F)(\ell^R_{\kappa,s} - \ell^L_{\kappa,s})$. Hence, we can set $F > 0$ without loss of generality. The discussion for $F < 0$ will only be reversed with respect to that of $F > 0$. Let us look at the same two cases as previously, starting with $\gamma = 0$. In this case, as we saw, the decaying lengths are independent of κ, which implies that the correction to the energy is just a constant offset. We already discussed that if $\delta \in (0, 1)$, then $\ell^R_{\kappa,s} < \ell^L_{\kappa,s}$, which in this case would imply that the Dirac cone will move downwards in energy. The opposite would happen if $\delta \in (-1, 0)$. In fact, an expansion around $\delta \to 0$ allows us to write the dispersion as follows

$$\varepsilon_{\kappa,s} \simeq \varepsilon^0_{\kappa,s} - f\delta. \tag{3.1.16}$$

Notice that, as we just said, it is the sign of δ or, equivalently, the sign of $(\ell^R_{\kappa,s} - \ell^L_{\kappa,s})$ (remember we have chosen $F > 0$), what will determine where the cone will move to. In Fig. 3.2, we show the shift $|\Delta\varepsilon| = |\varepsilon_{\kappa,s} - \varepsilon^0_{\kappa,s}|$, obtained from the exact result given by Eq. (3.1.15) and this last approximation for a value of $\delta = 0.2$ and f ranging

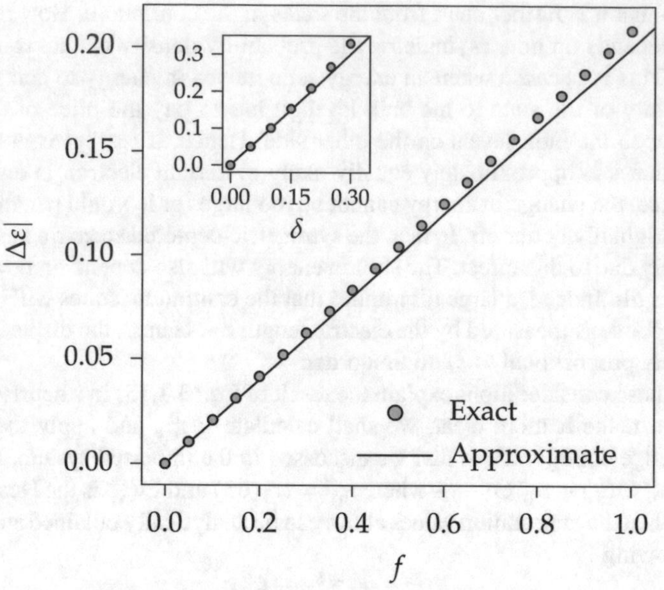

Fig. 3.2 Shift of the Dirac cone, $|\Delta\varepsilon| = |\varepsilon_{\kappa,s} - \varepsilon^0_{\kappa,s}|$, as a function of f for $\delta = 0.2$. The inset shows also the shift in the vertical axis, as a function of δ for $f = 1$. Dots correspond to the exact result (3.1.15) and solid lines to the approximation (3.1.16)

from 0 to 1. In the inset, f is fixed to 1 and δ is varied. In both cases $\gamma = 0$. As we can see, the agreement is noteworthy.

Let us try to understand this result, considering $\delta > 0$, the discussion being the exact opposite for $\delta < 0$. For that matter, imagine the tilting of the band edges due to the electric field potential, as shown in Fig. 3.3. As we can observe, if $\delta > 0$, upon tilting the band edges the bulk states on the left become closer to the Dirac point than those to the right. This means that the system would evolve so as to reduce its proximity to the bulk states on the left, which implies a reduction in energy. We can obtain this reduction heuristically by considering the crossing of the tilted band edges with the zero energy line, where the Dirac point sits in the unperturbed system. On the left, the conduction band-edge on the left would cross at $-\ell_F/\ell^L$ and the valence band-edge on the right would do so at ℓ_F/ℓ^R. The distance from the Dirac point to the valence band edge on the left would then be ℓ_F/ℓ^L, which is smaller than the distance to the right, ℓ_F/ℓ^R, as we anticipated. Since ℓ_L is already larger than ℓ_R, the Dirac point moves downwards in energy in order to stay away from the bulk states and remain protected. To lowest order in δ, we do so by reaching a compromise where the distance to the left is increased by δ and decreased on the right, also by δ. This is attained by moving the Dirac point downwards in energy by $-f\delta$. Since the decay lengths are independent of κ, all states will suffer the same fate as the Dirac point and the whole Dirac cone will only be displaced downwards in energy. Notice, however, that in contrast to the unperturbed situation, states of

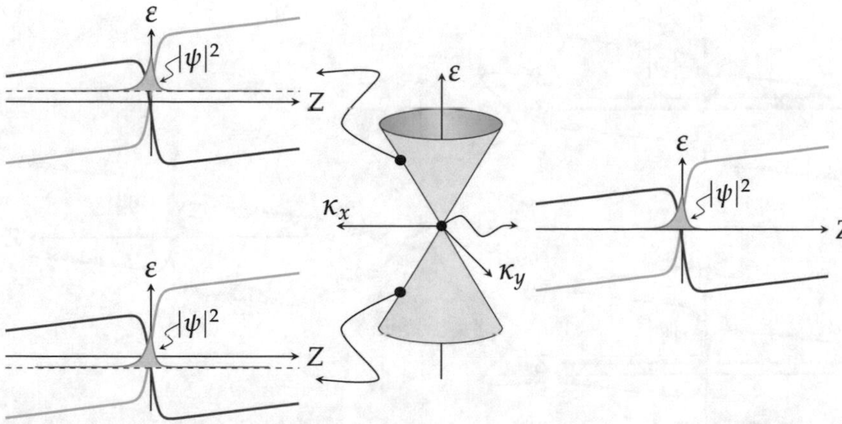

Fig. 3.3 Schematic decay of surface states in a centered, asymmetric topological boundary for different momenta. Band edges tilt due to the electric field potential. Surface states decay faster on the largest-gap side, independent of momentum. Above and below the band edges, there is a continuum of states

higher momenta would be closer to the continuum on the left (positive energies) and on the right (negative energies) than states of lower momenta, so the former should be affected by the field more severely. However, since the first order approximation considers the states as decaying just like if there was no perturbation and, in the case of centered gaps, the decay length is independent of κ, there can be no κ-dependent correction. This is a flaw of this first order approximation and shall be fixed in the next section. As we shall see, a correction that goes with f^2 will be κ-dependent. However, we shall observe when studying the exact solution to the problem that first order perturbation theory correctly predicts a global offset to the energy given by $f\delta$.

Let us now explore the situation where $\delta = 0$ and $\gamma \neq 0$. As we explained above, the decay lengths for the Dirac point in this case are the same on both sides. Thus, since the crossing of the band edges with the zero energy line occurs at exactly the same distance, $(1 - \gamma)/f$, the Dirac point remains in place. However, in contrast to the previous situation, these decay lengths do depend on κ in the manner described above. That is, if $\gamma > 0$, states above the Dirac point will decay slower to the left than to the right, and the opposite for those states below the Dirac point. In fact, the larger the momenta is, the stronger this effect turns out to be, see the discussion above. Upon introducing the electric field, the band edges tilt, as depicted schematically in Fig. 3.4.

States above zero energy move downwards in energy so as to decrease the decay length to the left. The same happens in the opposite direction for those states below zero energy. Since in-plane momentum is conserved in this scenario as well, the net result is, again, to reduce the Fermi velocity. Notice that there is an asymmetry when the field is introduced, in contrast to the field-free case. Indeed, now the effect

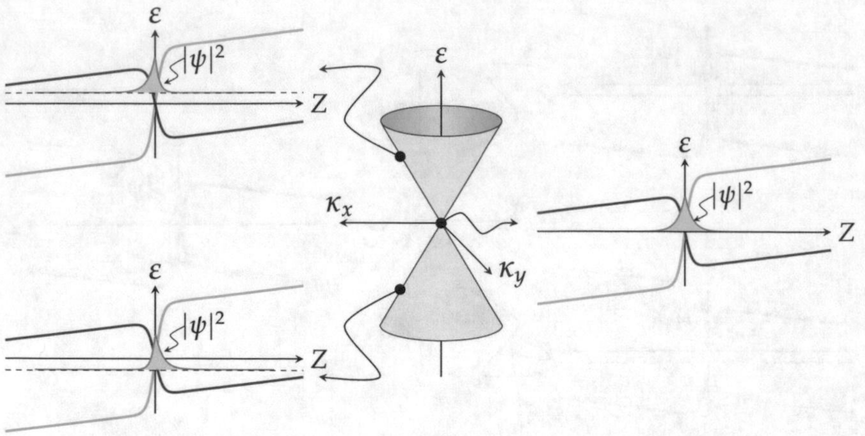

Fig. 3.4 Schematic decay of surface states in a symmetric off-centered topological boundary for different momenta. Band edges tilt due to the electric field potential. Surface states decay faster (slower) to the right the higher (lower) we go in energy with respect to the decay to the left. Above and below the band edges, there is a continuum of states

depends on γ and not on γ^2. More precisely, it depends on $f\gamma$, since changing the sign of both γ and f simultaneously would render the same situation, only inverting the Z-axis. Therefore, if $\gamma < 0$ the opposite to having $\gamma > 0$ happens, since now the conduction band edge on the right upon tilting is further up and the valence band edge on the left is further down. Although the opposite happens to the conduction and valence band edges on the left and right, respectively, these are further away in energy and, at least for low momenta, the situation is the one we just described. In turn, the Fermi velocity increases. In fact, a series expansion around $\gamma \to 0$ and $\kappa \to 0$ renders the following expression for the dispersion relation

$$\varepsilon_{\kappa,s} \simeq \varepsilon^0_{\kappa,s}\,(1 - f\gamma)\,. \tag{3.1.17}$$

As explained, if $f\gamma > 0$, the Fermi velocity decreases. Rather, if $f\gamma < 0$, it would increase. In Fig. 3.5a, we show the change in the Fermi velocity, $v_F(f)$ as a function of the field f for a fixed $\gamma = 0.3$, as obtained from the exact result in (3.1.15) and the approximation (3.1.17). Velocities from the exact result have been obtained from linear fits of the dispersion up to $\kappa = 0.5$, although it remains linear even up to $\kappa = 1$, with only about 5% difference in the ratio $\varepsilon(\kappa)/\kappa$ between these two values of κ for $f = 1$. The inset shows the case where $f = 1$ and γ is varied. In both cases $\delta = 0$ and the agreement is significant.

Similarly, Fig. 3.5b shows the exact same situation, except for having a negative γ. As can be observed, the Fermi velocity increases in this case, as discussed. In fact, as we already said, the results would be identical upon changing $f \to -f$ in the second figure and extending the regime to $f \in [-1, 1]$.

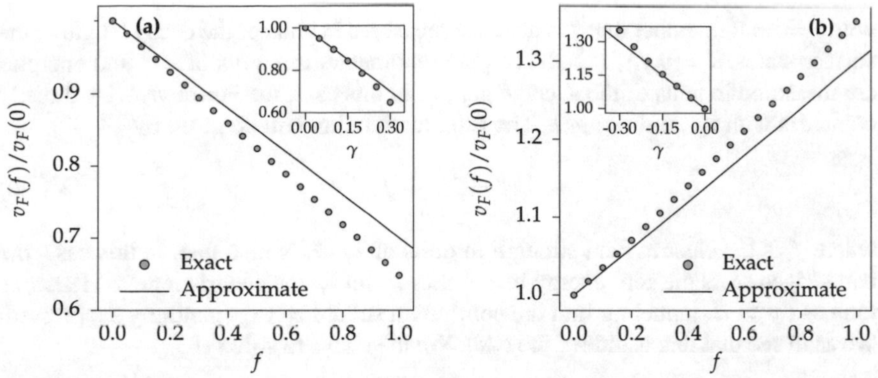

Fig. 3.5 Fermi velocity $v_F(f)$ as a function of f. a $\gamma = 0.3$ and **b** $\gamma = -0.3$. $v_F(0)$ corresponds to the field-free velocity. The inset shows the velocity in the vertical axis as well, now as a function of γ for $f = 1$. Dots correspond to the exact result (3.1.15) and solid lines to the approximation (3.1.17)

Notice that the arguments given herein can only hold for small values of the field and momenta. Indeed, if the field is too large, even though the states will try to minimize the effect by reshaping the cone, there will come a point where the field will be such that the tilting becomes sufficiently close so that the surface states will unavoidably leak into the continuum. In fact, this situation is what happens in reality, as we shall see next, and states become *resonant* or *quasi-bound* states, since they acquire a finite lifetime due to such leakage. However, we shall observe that this decay into the continuum is rather small for small enough fields and, therefore, it can be neglected in practice, at least for low momenta.

3.1.2 Approximate Solution

In the last section, we could observe that first order perturbation theory predicted that, for centered gaps, the correction to the dispersion was to displace the cone in energy in order to minimize the effect of the field. In this section, we will present an approximate solution to the problem of centered gaps. Since we already know that the first order correction predicts a displacement of the Dirac point, we shall focus on same-sized gaps in order to isolate the effect of the field on structures with centered gaps. Hence, the unperturbed Hamiltonian reads

$$\mathcal{H}_0 = -\mathrm{i}\,\alpha_z\partial_\xi + \boldsymbol{\alpha}_\perp\cdot\boldsymbol{\kappa} + \mathrm{sgn}\,(\xi)\beta. \tag{3.1.18}$$

In this case, the dispersion is simply

$$\varepsilon_{\kappa,s} = s\kappa, \tag{3.1.19}$$

with $s = \pm$. Remember that ξ is distance measured in units of the decay length of the surface states, $d = v_z/\lambda$, $\boldsymbol{\kappa}$ is the in-plane momentum in units of d^{-1} and energies are measured in units of the average gap, λ. In this case, the Fermi velocity directly coincides with v_z, so $d = v_F/\lambda$. The full Hamiltonian will be given by

$$\mathcal{H} = \mathcal{H}_0 + fz, \tag{3.1.20}$$

where f is the electric field strength in units of λ/ed. Notice that, in this case, the band edges cross the zero energy line at $\pm\ell_F$, with ℓ_F the electric length. Hence, as long as $\ell_F \gg d$, tunneling into the continuum should be exponentially suppressed. We shall see that this is indeed the case. We then have to solve

$$\mathcal{H}_0 \boldsymbol{\chi}(\xi) = (\varepsilon - f\xi)\, \boldsymbol{\chi}(\xi). \tag{3.1.21}$$

As is customary with the Dirac equation, we apply $\mathcal{H}_0$ again on the left

$$\mathcal{H}_0^2 \boldsymbol{\chi}(\xi) = \varepsilon(\varepsilon - f\xi)\boldsymbol{\chi}(\xi) - f\,(\mathcal{H}_0\xi)\, \boldsymbol{\chi}(\xi). \tag{3.1.22}$$

We can write $\mathcal{H}_0\xi$ as follows

$$\mathcal{H}_0\xi = [\mathcal{H}_0, \xi] + \xi\mathcal{H}_0, \tag{3.1.23}$$

where $[A, B] = AB - BA$ is the commutator. Since $[\partial_\xi, \xi] = 1$, we can see that

$$\mathcal{H}_0\xi = -\mathrm{i}\,\alpha_z + \xi\mathcal{H}_0. \tag{3.1.24}$$

Hence, Eq. (3.1.22) can be written as follows

$$\mathcal{H}_0^2 \boldsymbol{\chi}(\xi) = \left[(\varepsilon - f\xi)^2 + \mathrm{i}\, f\alpha_z\right] \boldsymbol{\chi}(\xi). \tag{3.1.25}$$

The term $\mathcal{H}_0^2$ is also easily evaluated taking into account that $\partial_z \mathrm{sgn}\,(z) = 2\delta(z)$. All in all, we can write the following equation for $\boldsymbol{\chi}(\xi)$

$$\left[-\frac{\mathrm{d}^2}{\mathrm{d}\xi^2} + U(\xi) - \mathrm{i}\, f\alpha_z + 2\varepsilon f\xi - f^2\xi^2 + \rho^2\right] \boldsymbol{\chi}(\xi) = 0, \tag{3.1.26}$$

where we have introduced

$$U(\xi) = 2\mathrm{i}\,\beta\alpha_z\delta(\xi), \qquad \rho^2 = 1 + \kappa^2 - \varepsilon^2. \tag{3.1.27}$$

Notice that ρ^2 is nothing but $\pi(\xi)$, which in the symmetric junction is independent of ξ. The approximate solution that we shall present here considers low fields, f. Equivalently, we are interested in the regime where the electric length, ℓ_F, is much larger than the decay length of the surface states, d. Since d is of the order of nanometers, as we discuss below, it is convenient to write ℓ_F as follows

$$\ell_F\,[\mathrm{nm}] = 10\frac{\lambda[\mathrm{meV}]}{eF[\mathrm{kV/cm}]}, \tag{3.1.28}$$

which gives the electric length in nanometers when λ is given in meV and F is given in kV/cm. Before we turn to real materials, it is interesting to have an estimate for an intermediate $\lambda \simeq 100$ meV. In this case, an electric length of $\ell_F \simeq 10$ nm would correspond to an electric field of $F \simeq 100$ kV cm^{-1}, which is not negligible for experiments. Let us consider real materials. In Bi_2Se_3, $\lambda \simeq 175$ meV [4] and $v_F \simeq 0.25$ eV nm [5], which leads to $d \simeq 1.43$ nm and a value of $F_C \simeq 1220$ kV cm^{-1}. On the other hand, typical values for narrow-gap IV-VI semiconductors [6] such as SnTe are $\lambda \simeq 75$ meV and $v_F \simeq 0.34$ eV nm, leading to $d \simeq 4.5$ nm and a much lower but still large value of the critical field of about $F_C \simeq 170$ kV cm^{-1}. Therefore, the approximation $|f| \ll 1$ is not only compelling from the theoretical point of view, but also from the experimental one. In this approximation, we can neglect the $f^2\xi^2$ term, arguing that the extent of the surface states is rather small in comparison to $1/f$, that is, in comparison the electric length. Regarding the term $-\mathrm{i}\, f\alpha_z$, it can be brought into a diagonal form and be neglected in comparison to ρ^2. The reason is that $\rho^2 \approx 1$ since for small enough fields the energies cannot deviate very much from the Dirac-like spectrum. In the section dedicated to the exact solution we shall see that this term can indeed be neglected. Therefore, we must solve the following simplified problem

$$\left[-\frac{\mathrm{d}^2}{\mathrm{d}\xi^2} + U(\xi) + 2\varepsilon f\xi + \rho^2\right]\chi(\xi) = 0. \tag{3.1.29}$$

This problem is very much like the one solved by Ludviksson of a particle in a tilted potential trapped by a δ well at the origin [7–9], with the difference that $U(\xi)$ is not a scalar here. We know from the field-free case that the topological boundary has two types of states: a continuum of states above and below the conduction and valence band edges, and a bound state localized at the boundary. In this problem, we still have a continuum of states above and below the tilted band edges. However, the bound states now become *quasi-bound* [7] or *resonant states*, as we discussed above. Indeed, due to the band tilting, the original bound state acquires a finite lifetime due to the possibility of tunneling into the continuum. Since we are considering centered and same-sized gaps, the discussion presented earlier would imply that the decay lengths of the bound states on both sides would be the same, irrespective also of κ. However, we already pointed out that that had to be a flaw of the first order approximation. In fact, our intuition would tell us that for energies above $\varepsilon = 0$ the resonant state will be more localized on the right of the boundary, because it will be closer to the continuum states on the left, and vice versa for energies below $\varepsilon = 0$. At $\varepsilon = 0$, however, it will be equally localized on both sides due to the symmetry of the problem and, if the field is not strong enough, the Dirac point should remain as if there was no field. In fact, at $\varepsilon = 0$, Eq. (3.1.29) becomes independent of the field and, therefore, $\varepsilon = 0$ for $\kappa = 0$ is still an eigenenergy when the field is applied. In other words, that state remains a stationary bound state. We shall see that this is indeed the case, when we discover that there is no gap opening at the Dirac point and that it has an infinite lifetime, a fact that is ultimately linked to the fact that time-reversal symmetry is preserved, as we discussed earlier in the text.

With this in mind, let us then continue solving this problem. In order to do so, it is convenient to regard the term $U(\xi)$ as a perturbation. The reason for this approach will become clear shortly. The perturbation-free problem can be solved within the Green function approach

$$\left[-\frac{\partial^2}{\partial\xi^2}+2\varepsilon f\xi+\rho^2\right]\mathcal{G}_0(\xi,\xi';\varepsilon)=\delta(\xi-\xi')\mathbb{1}_4. \tag{3.1.30}$$

Since the operator acting upon $\mathcal{G}_0(\xi,\xi';\varepsilon)$ is a scalar, we can factorize the Green function as $\mathcal{G}_0(\xi,\xi';\varepsilon)=G_0(\xi,\xi';\varepsilon)\mathbb{1}_4$. Once $\mathcal{G}_0(\xi,\xi';\varepsilon)$ is known, we can proceed to use the homogeneous *Lippman–Schwinger* equation [10] to obtain $\boldsymbol{\chi}(\xi)$ for the quasi-bound state. Notice that we can use the homogeneous equation because we are interested in the quasi-bound states, which do not belong to the spectrum of the perturbation-free problem, for otherwise they would be bound states. We will comment on this statement later on. As of now, it suffices to say that the Lippman–Schwinger equation allows us to obtain $\boldsymbol{\chi}(\xi)$ as follows

$$\boldsymbol{\chi}(\xi)=-\int \mathrm{d}\xi'\mathcal{G}_0(\xi,\xi';\varepsilon)U(\xi')\boldsymbol{\chi}(\xi'). \tag{3.1.31}$$

Generally, the form of $U(\xi)$ does not allow for an exact solution to this integral equation and it has to be addressed perturbatively. However, the functional form of $U(\xi)$ in our present case allows us to perform the integral and we obtain

$$\boldsymbol{\chi}(\xi)=-\mathcal{G}_0(\xi,0;\varepsilon)2\mathrm{i}\,\beta\alpha_z\boldsymbol{\chi}(0). \tag{3.1.32}$$

Taking into account that $\mathcal{G}_0(\xi,\xi';\varepsilon)$ is proportional to the identity matrix, we can write for the probability density

$$|\boldsymbol{\chi}(\xi)|^2=4|G_0(\xi,0;\varepsilon)|^2|\boldsymbol{\chi}(0)|^2, \tag{3.1.33}$$

where $|\boldsymbol{\chi}(\xi)|^2=\boldsymbol{\chi}^\dagger(\xi)\boldsymbol{\chi}(\xi)$. Therefore, the probability density profile can be studied by studying the absolute squared value of $G_0(\xi,0;\varepsilon)$. Both Eqs. (3.1.32) and (3.1.33) have to hold for all values of ξ and, in particular, they must hold for $\xi=0$. From Eq. (3.1.33) we find that in order to have nontrivial solutions, the following must be required

$$4|G_0(0,0;\varepsilon)|^2=1. \tag{3.1.34}$$

This equation has to be solved for ε and it would provide us with the energies of the quasi-bound states. However, it is also interesting to take a look at Eq. (3.1.32) when particularized for $\xi=0$

$$\chi(0) = -\mathcal{G}_0(0,0;\varepsilon)2\mathrm{i}\,\beta\alpha_z\chi(0). \tag{3.1.35}$$

The meaning of this equation is the following: we have to find those values of ε for which the eigenvalues of $-\mathcal{G}_0(0,0;\varepsilon)2\mathrm{i}\,\beta\alpha_z$ are equal to one. The eigenvalues of such a matrix, taking into account that $\mathcal{G}_0(0,0;\varepsilon)$ is proportional to the identity matrix, are given by

$$\gamma_G = \pm 2G_0(0,0;\varepsilon), \tag{3.1.36}$$

each of which is doubly degenerate. The requirement that $\gamma_G = 1$ implies that

$$1 = \pm 2G_0(0,0;\varepsilon)\,, \tag{3.1.37}$$

which, if taken the squared absolute value, leads to Eq. (3.1.34). However, this equation tells us more than Eq. (3.1.34). Indeed, it is telling us that ε has to be within a range where $G_0(0,0;\varepsilon)$ is real. Not only that, it is also telling us that the absolute value in Eq. (3.1.34) can be removed and the square suffices. We shall see the consequences of this shortly. This equation is also telling us that if $G_0(0,0;\varepsilon)$ is positive (negative) we can rule out the two other eigenvectors. The similarity with the fact that we could get rid of two eigenvectors when calculating the surface states in the second chapter is not accidental. Indeed, the condition we just found is totally equivalent to that found in the previous chapter.

Before we move on to obtaining $G_0(\xi,\xi';\varepsilon)$ for the electric field problem, let us ponder on the results thus far and how would they apply for the field-free problem. As they are, Eq. (3.1.35) and those thereafter apply to the field-free case as well. Therefore, we should be able to obtain the same results that we obtained in the previous chapter. Indeed we do, because the Green function in that case can be shown to be [10]

$$G_0(\xi,\xi';\varepsilon) = \frac{1}{2\rho}\exp(-\rho|\xi-\xi'|), \tag{3.1.38}$$

where $\rho > 0$ is still given by Eq. (3.1.27). From Eq. (3.1.37) to hold we must ask for $\rho^2 = 1$, which in turn leads to the Dirac cone, $\varepsilon = \pm\kappa$. Also, $G(0,0;\varepsilon)$ is positive, so we can consider only those eigenvectors that correspond to the positive eigenvalue of one. Those eigenvectors are the Kramers' pairs we obtained in the previous chapter.

Let us now turn our attention to the electric field problem. We have to solve Eq. (3.1.30), subject to the boundary conditions that $G_0(\xi,\xi';\varepsilon)$ is bounded as $|\xi| \to \infty$. From our previous discussion, we saw that $\varepsilon = 0$ is still an eigenenergy of the system for $\kappa = 0$. Let us then focus on the problem for $\varepsilon \neq 0$. We can make the assumption that, for low energies, the imaginary part of ε, which provides information about the lifetime of a given state, is vanishingly small. In other words, if we write $\varepsilon = |\varepsilon|\exp(\mathrm{i}\,\theta_\varepsilon)$, the fact that the imaginary part is small allows us to write $\varepsilon \approx s_\varepsilon|\varepsilon|$, where $s_\varepsilon = \mathrm{sgn}\,[\mathrm{Re}(\varepsilon)]$. We shall do this approximation only for the factor that accompanies ξ in Eq. (3.1.30). That is, when applying the boundary conditions we will take into account the fact that ρ^2 is complex. Let us introduce then the following definitions

$$\mu = (2|\varepsilon|f)^{1/3}, \qquad \eta = s_\varepsilon \mu \xi + \frac{\rho^2}{\mu^2}, \qquad \widetilde{G}_0 = \mu G_0. \tag{3.1.39}$$

Then, Eq. (3.1.30) may be written as follows

$$\left[-\frac{\partial^2}{\partial \eta^2} + \eta\right] \widetilde{G}_0(\eta, \eta'; \varepsilon) = \delta(\eta - \eta'). \tag{3.1.40}$$

When $\eta \neq \eta'$, this equation corresponds to the Airy equation [11], so two independent solutions are given by the Airy functions $\mathrm{Ai}\,(\eta)$ and $\mathrm{Bi}\,(\eta)$. In order to find the Green function, we must make linear combinations of these two such that they satisfy the boundary conditions. On the one hand, $\mathrm{Ai}\,(\eta)$ satisfies the boundary condition at $\eta \to \infty$, for any value of $\mathrm{Im}(\rho^2)$. On the other hand, $\mathrm{Ci}^{\pm}(\eta) = \mathrm{Bi}\,(\eta) \pm \mathrm{i}\,\mathrm{Ai}\,(\eta)$ satisfy the boundary condition at $\eta \to -\infty$ if $\mathrm{sgn}\left[\mathrm{Im}(\rho^2)\right] = \mp 1$, the upper sign corresponding to $\mathrm{Ci}^{+}(\eta)$ and the lower sign to $\mathrm{Ci}^{-}(\eta)$. Let $s_\rho = -\mathrm{sgn}\left[\mathrm{Im}(\rho^2)\right]$. Then, we can write $\mathrm{Ci}^{s_\rho}(\eta) = \mathrm{Bi}\,(\eta) + s_\rho \mathrm{i}\,\mathrm{Ai}\,(\eta)$. As a result, the Green function is given by

$$\widetilde{G}_0^{s_\rho}(\eta, \eta'; \varepsilon) = \frac{1}{\mathcal{W}\left[\mathrm{Ci}^{s_\rho}(\eta), \mathrm{Ai}\,(\eta)\right]} \begin{cases} \mathrm{Ci}^{s_\rho}(\eta')\mathrm{Ai}\,(\eta), & \text{if } \eta' \leq \eta, \\ \mathrm{Ci}^{s_\rho}(\eta)\mathrm{Ai}\,(\eta'), & \text{if } \eta' \geq \eta, \end{cases} \tag{3.1.41}$$

where we have made explicit the dependence of s_ρ on $\tilde{G}$ and we have denoted the Wronskian as $\mathcal{W}\left[f(\eta), g(\eta)\right] = f(\eta)\partial_\eta g(\eta) - g(\eta)\partial_\eta f(\eta)$. In this case, the Wronskian is simply [11]

$$\mathcal{W}\left[\mathrm{Ci}^{s_\rho}(\eta), \mathrm{Ai}\,(\eta)\right] = -\frac{1}{\pi}. \tag{3.1.42}$$

Therefore, the Green function is given by

$$\begin{aligned} \widetilde{G}_0^{s_\rho}(\eta, \eta'; \varepsilon) = -\pi \Big[& \Theta(\eta - \eta')\mathrm{Ci}^{s_\lambda}(\eta')\mathrm{Ai}\,(\eta) \\ & + \Theta(\eta' - \eta)\mathrm{Ci}^{s_\rho}(\eta)\mathrm{Ai}\,(\eta') \Big], \end{aligned} \tag{3.1.43}$$

where $\Theta(x)$ is the Heaviside step function. Undoing the change of variables defined in (3.1.39), we can write

$$\begin{aligned} G_0^{s_\rho}(\xi, \xi'; \varepsilon) = -\frac{\pi}{\mu} \Big\{ & \Theta\left[s_\varepsilon(\xi - \xi')\right] \mathrm{Ai}\,(\eta(\xi))\mathrm{Ci}^{s_\rho}(\eta(\xi')) \\ & + \Theta\left[s_\varepsilon(\xi' - \xi)\right] \mathrm{Ai}\,(\eta(\xi'))\mathrm{Ci}^{s_\rho}(\eta(\xi)) \Big], \end{aligned} \tag{3.1.44}$$

where $\eta(\xi)$ is given in Eq. (3.1.39). Let us discuss these results before we move on. In Eq. (3.1.44) we have shifted from a single $G_0(\xi, \xi'; \varepsilon)$ to two options that depend on s_ρ. Therefore, there seems to be an inconsistency in our calculations. Indeed, Eq. (3.1.31) truly holds if we are dealing with bound states, when the spectrum of the continuum and that of the bound states do not coincide. However, in our case, these two spectra do coincide due to the band tilting, which allows an electron in a quasi-bound state to tunnel into the continuum. This is of course the reason for their being called quasi-bound or resonant states. However, if the fields are small enough, an approximation that we have already discussed to be sensible, this leakage into the continuum is rather small, at least within a narrow energy window around zero energy. That is, quasi-bound states are almost pure bound states. In that case, we can forget about the imaginary part of ε and both G_0^+ and G_0^- would be equivalent by substituting $\mathrm{Ci}^{\pm}(\xi)$ by $\mathrm{Bi}(\xi)$. However, it is interesting to obtain the imaginary part of ε to show that, indeed, it is rather small and, therefore, we are making a good approximation when considering that quasi-bound states are almost bound states. There is an alternative way to approach this problem, which is by considering the total Green function of the problem via the *Dyson equation*, as described in [1]. In that reference, it is shown that the results are totally equivalent to the ones presented in this section. Having said this, there is yet another point to notice. Equation (3.1.34), a consequence of (3.1.33), can only lead to real energies. Indeed, it derives from the fact that the equation it comes from is for true bound states, with no imaginary part in the energy. However, if we take into account Eq. (3.1.37) and we square it, we end up with a less strict condition as that given in Eq. (3.1.34)

$$4\,[G_0(0,0;\varepsilon)]^2 = 1. \tag{3.1.45}$$

Now we may raise an objection. Indeed, this equation also comes from the homogeneous Lippman–Schwinger equation, so it must lead to the same results as those from (3.1.34). This is in fact the case, when we consider that there is no imaginary part to the energy and $G_0^+ = G_0^-$. However, it turns out that this precise equation comes about when studying the poles of the total Green function when we allow for an imaginary part in the energy. That is, Eq. (3.1.45) also holds for $G_0^{\pm}$ [1]. As said earlier, for small enough fields, the results obtained from using such an equation with G_0 or $G_0^{\pm}$ are essentially the same. Indeed, when $f \ll 1$, we can approximate $\mathrm{Ci}^{s_\rho}(\eta(0)) \simeq \mathrm{Bi}(\eta(0))$. The reason for this is that, at low fields, $\rho^2 \simeq 1$, because the dispersion should not change very much from being a Dirac cone as the field is adiabatically turned on. Hence, the quotient ρ^2/μ^2 increases with decreasing f, meaning that the argument $\eta(0)$ increases with decreasing f. Due to the properties of the Airy functions, this means that $\mathrm{Ci}^{s_\rho}(\eta(0)) \simeq \mathrm{Bi}(\eta(0))$. Nevertheless, it is interesting to use that equation with either G_0^+ or G_0^- to obtain the level width and observe that, indeed, it is truly small for small fields. Hence, we may focus on the following equation to obtain analytical results at low fields

$$\frac{4\pi^2}{\mu^2}\left[\mathrm{Ai}\left(\frac{\rho^2}{\mu^2}\right)\mathrm{Ci}^{s_\rho}\left(\frac{\rho^2}{\mu^2}\right)\right]^2 = 1. \tag{3.1.46}$$

There is one crucial fact here: $G_0^{s_\rho}$ does not satisfy the boundary conditions for $-s_\rho$, which is why it has to be separated depending on s_ρ in the first place. As a result, the simple poles of the total Green function appear when we analytically continue from s_ρ to $-s_\rho$. Consequently, the previous equation, that provides the poles of the total Green function [1], holds when studying the regime where we have $-s_\rho$ for a given s_ρ in the superscript of $\mathrm{Ci}^{s_\rho}(\xi)$. In order to make this clearer, consider the previous equation for a particular choice of s_ρ in the superscript of Ci, say $s_\rho = 1$. Then, Eq. (3.1.46) reads

$$\frac{4\pi^2}{\mu^2}\left[\mathrm{Ai}\left(\frac{\rho^2}{\mu^2}\right)\mathrm{Ci}^+\left(\frac{\rho^2}{\mu^2}\right)\right]^2 = 1. \tag{3.1.47}$$

The comments that we have made prior to this equation mean that the solutions to this equation will be such that $s_\rho = -1$, that is, $\mathrm{sgn}\left[\mathrm{Im}(\rho^2)\right] > 0$. In other words, it will have that the real and imaginary parts of ε will satisfy that $\mathrm{Re}(\varepsilon)\mathrm{Im}(\varepsilon) < 0$. Hence, if we let $\Gamma > 0$ be the imaginary part of ε, we will be able to write $\varepsilon = \varepsilon_r \pm \mathrm{i}\,\Gamma$, where the positive sign stands for $\varepsilon_r < 0$ and the negative sign for $\varepsilon_r > 0$. The opposite is true if we choose instead Ci^- in the previous equation. With this in mind, we can push further the low-field limit by making asymptotic expansions around $f \to 0$. For that matter, let us rewrite Eq. (3.1.47) as follows

$$\frac{4\pi^2}{\mu^2}\mathrm{Ai}^2(x)\mathrm{Bi}^2(x)\left[1 + \mathrm{i}\,\frac{\mathrm{Ai}\,(x)}{\mathrm{Bi}\,(x)}\right]^2 = 1, \tag{3.1.48}$$

where we have introduced

$$x = \frac{\rho^2}{\mu^2}. \tag{3.1.49}$$

We will also introduce the asymptotic expansions [11] of Ai (x) and Bi (x) as $x \to \infty$, which corresponds in our case to $f \to 0$ (recall that, in that limit, ρ^2 remains finite and close to 1),

$$\mathrm{Ai}\,(x) \simeq \frac{1}{2\sqrt{\pi}}\frac{e^{-\phi}}{x^{1/4}}L(-\phi), \qquad \mathrm{Bi}\,(x) \simeq \frac{1}{\sqrt{\pi}}\frac{e^{\phi}}{x^{1/4}}L(\phi), \tag{3.1.50}$$

where

$$\phi = \frac{2}{3}x^{3/2}, \qquad L(\phi) = 1 + \sum_{\ell=1}^{\infty}\frac{u_\ell}{\phi^\ell}, \qquad u_\ell = \frac{\Gamma(3\ell + 1/2)}{54^\ell l!\,\Gamma(\ell + 1/2)}, \tag{3.1.51}$$

being $\Gamma(x)$ the Γ function. Taking into account that Ai $(x) \ll$ Bi (x) if $x \to \infty$, Eq. (3.1.48) can be approximated to

$$\frac{4\pi^2}{\mu^2}\mathrm{Ai}^2(x)\mathrm{Bi}^2(x)\left[1 + 2\mathrm{i}\,\frac{\mathrm{Ai}\,(x)}{\mathrm{Bi}\,(x)}\right] \simeq 1. \tag{3.1.52}$$

If the asymptotic expansions are taken into account, we see that the previous equation can be rewritten as follows

$$\rho^2 \simeq L^2(\phi)L^2(-\phi)\left[1+\mathrm{i}\,e^{-2\phi}\frac{L(-\phi)}{L(\phi)}\right]. \tag{3.1.53}$$

We can further approximate this equation by keeping only terms up to ϕ^{-2}

$$\rho^2 \simeq \left[1+\frac{5}{36\phi^2}\right]\left[1+\mathrm{i}\,e^{-2\phi}\right] \simeq 1+\frac{5}{36\phi^2}+\mathrm{i}\,e^{-2\phi}. \tag{3.1.54}$$

Taking into account the definition of ϕ, we can finally write

$$\rho^2 \simeq 1+\frac{5}{4}\frac{(|\varepsilon|\,f)^2}{\rho^6}+\mathrm{i}\,\exp\left(-\frac{4}{3}\frac{\rho^3}{\mu^3}\right). \tag{3.1.55}$$

At this stage, we have to make a few more approximations. First, since for low fields the imaginary part of the energy should be small, then we will approximate $|\varepsilon| \simeq |\varepsilon_r|$. On the other hand, as we stated above, $\rho^2 \simeq 1$, so we shall approximate to one the factor of ρ^6 in the denominator of the expression above and ρ^3 in the exponential. Finally, in the expression for ρ^2, we will approximate $\varepsilon^2 \simeq \varepsilon_r^2 \pm 2\mathrm{i}\,\varepsilon_r\Gamma$, where the positive sign stands for $\varepsilon_r < 0$ and the negative sign for $\varepsilon_r > 0$, as explained above. With all these considerations, we find that we can write

$$\kappa^2-\varepsilon_r^2 \mp 2\mathrm{i}\,\varepsilon_r\Gamma \simeq \frac{5}{4}\varepsilon_r^2 f^2+\mathrm{i}\,\exp\left(-\frac{2}{3|\varepsilon_r|\,f}\right). \tag{3.1.56}$$

Identically,

$$\varepsilon_r \simeq \pm\kappa\left(1+\frac{5}{4}f^2\right)^{-1/2}, \qquad \Gamma \simeq \mp\frac{1}{2\varepsilon_r}\exp\left(-\frac{2}{3|\varepsilon_r|\,f}\right). \tag{3.1.57}$$

Notice that this is consistent. Indeed, $\Gamma > 0$ always, because $\varepsilon_r < 0$ corresponds to the upper, negative sign, whereas $\varepsilon_r > 0$ corresponds to the lower, positive sign. On the other hand, the low field limit can be carried even further, allowing us to write

$$\varepsilon_r \simeq \pm\kappa\left(1-\frac{5}{8}f^2\right), \qquad \Gamma \simeq \frac{1}{2\kappa}\exp\left(-\frac{2}{3\kappa f}\right). \tag{3.1.58}$$

These are the main results of this section. On the one hand, we can observe that there is no gap opening, as expected since Kramers' degeneracy protects the Dirac point, but also because at $\kappa = 0$, the state remained an eigenstate of $\varepsilon = 0$. That is further supported by the fact that Γ tends to zero at $\kappa = 0$, meaning that the zero energy level has zero width or, equivalently, infinite lifetime. This in turn implies what we said earlier in the text: the Dirac point states are still pure bound states. The other result that is most interesting is the fact that the Fermi velocity decreases with increasing field

$$\frac{v_F(f)}{v_F(0)} = 1 - \frac{5}{8} f^2, \tag{3.1.59}$$

where $v_F(0)$ is the Fermi velocity in the absence of the electric field. This is precisely the result we were looking for in the previous section. Indeed, on the one hand we observe that there are no terms linear in the field, as it should be since we already ruled out those from first order perturbation theory in the symmetric setup. On the other, it leads to the increased reduction in the energy for states of larger momenta, which was absent from the analysis of the previous section for centered gaps. We also discussed that the Dirac state is sharp, that it has an infinite lifetime, but it is also interesting to observe that the level width increases with κ (recall that κ is small, we are dealing with a low energy description), which is related to the fact that states that are located higher (lower) in energy with respect to the Dirac cone are also closer to the conduction (valence) band edge on the left (right) and can therefore leak into the continuum more easily. In Fig. 3.6, we show the dispersion for two values of the field, dots being the numerical solution to Eq. (3.1.47), solid lines to the approximated dispersion as given in Eq. (3.1.58). As it can be drawn from the figure, by increasing the field the cone widens and, as the field gets closer to the critical field only energies of lower momenta keep a linear behaviour and deviate from the approximated result.

In Fig. 3.7a, it is shown the Fermi velocity as a function of the electric field strength as obtained by fitting the dispersion to a line up to $\kappa = 0.2$. The agreement with the approximated results at small fields is noteworthy. In Fig. 3.7b we show the level width as a function of the field for two values of momenta, dots being the numerical solution to Eq. (3.1.47), solid lines corresponding to the approximation in Eq. (3.1.58). As was predicted, the level width increases with the field and is also larger for larger momenta since these are closer to the band edges. In both cases, however, it is exponentially small for low fields, as discussed.

Before finishing up this section, we conclude by recapitulating the main physics. In the unbiased boundary, the Dirac point is protected by time-reversal symmetry. Away from the Dirac point, states are not Kramers' degenerate and split forming a Dirac cone. When we introduce the electric field, time-reversal symmetry is preserved and the Dirac point remains protected. However, the band edges tilt, thereby allowing bulk states to come closer to states at the surface, this effect being more pronounced for those states of higher momenta. In order to minimize the overlap with the bulk states, all states above and below the Dirac point move towards it. States of higher momenta experience a larger displacement than those of lower momenta for the reason we just stated. However, in-plane momentum conservation implies that a state of a given momentum remains at that same momentum. Therefore, states moving towards the Dirac cone have to do it in such a way as to preserve that conservation. All in all, the resulting effect is a widening of the Dirac cone, thereby effectively reducing the Fermi velocity. This is shown schematically in Fig. 3.8.

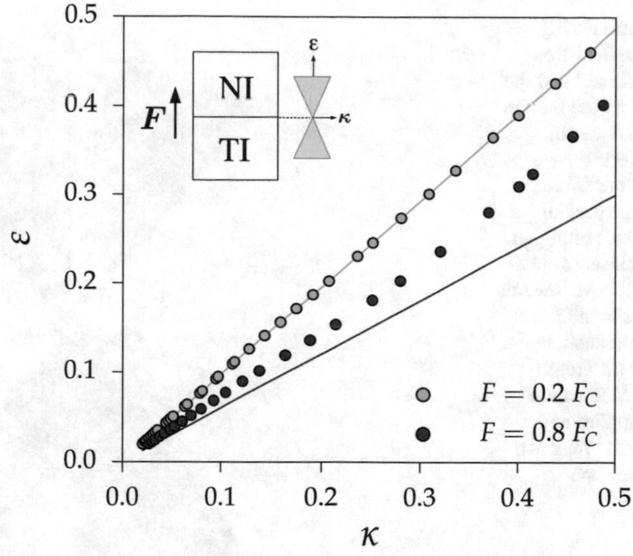

Fig. 3.6 Dispersion relation of a topological boundary under a perpendicular electric field. Two values of the electric field are shown. Solid lines correspond to the approximate result (3.1.58). Dispersion grows faster than linear only at high fields. The inset shows the electric field being applied perpendicular to the boundary between a normal insulator (NI) and topological insulator (TI), where the Dirac cones live

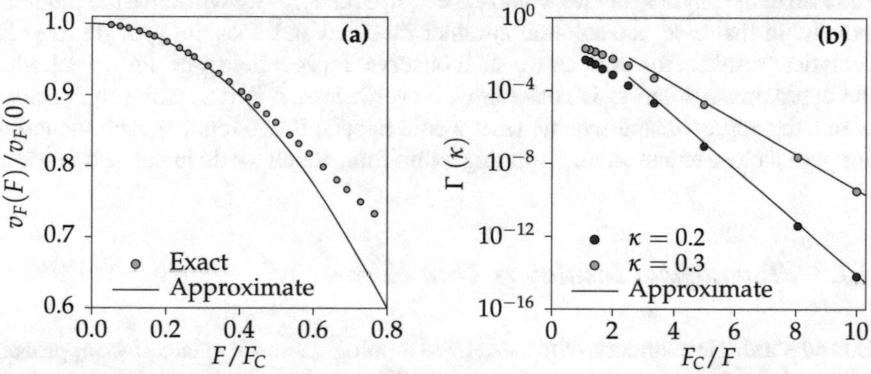

Fig. 3.7 Fermi velocity and level width as a function of electric field strength. **a** Fermi velocity obtained by fitting the numerical dispersions to a line up to $\kappa = 0.2$. Solid line displays the approximated reduction of the velocity as predicted by (3.1.59). **b** Level width as a function of the inverse of the electric field for two values of the in-plane momenta. Solid lines depict the approximation (3.1.58)

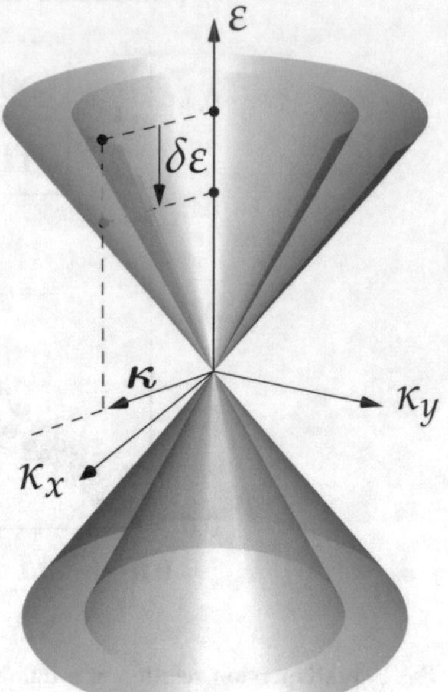

Fig. 3.8 Fermi velocity reduction. The field-free cone is shown in red and the cone reshaped by the field in blue. The field does not break time-reversal symmetry and the Dirac point remains protected. Meanwhile, states of higher momenta are closer to states in the bulk and move towards the Dirac point to minimize the effect. The reduction, $\delta\varepsilon$, is different for different momenta, but momentum conservation applies and, therefore, the net effect is a widening of the cone

Although all the approximations that have been discussed thus far seem reasonable and lead to the physics that we would expect, in Sect. 3.1.4 we will solve the problem exactly. In that case, the solution is rather involved and does not provide us with analytical results. However, as we shall observe, for low fields the agreement with the approximate solution is noteworthy. Nevertheless, it is interesting to consider within the approximated scheme what would happen if two topological boundaries are placed close to one another forming a thin film, which we do in the next section.

3.1.3 Topological Insulators Thin Film

Consider a single boundary, with a localized topological surface state. If we approach another boundary of the same type, there will be a gap opening and the massless fermions would become massive and doubly degenerate. Let us first try to understand this by thinking about the decay of the surface states out of the surface. If the two boundaries are sufficiently close, the overlap between surface states will not be negligible. Therefore, the two states hybridize as in the H_2^+ molecule and a gap opens up. Since each boundary contributes with a singly degenerate Dirac cone, the result is a doubly degenerate massive Dirac spectrum. However, this phenomenon is only relevant if the two surfaces are really close, since the decay length of the surface states is of the order of a few nanometers. Hence, this behaviour is important in thin films of topological insulators [12–18]. We can also understand the gap opening from a topological point of view. Indeed, consider an interface where the gap changes from $-\lambda$ to λ, with $\lambda > 0$. This case corresponds to the one discussed in the previous chapter and the Hamiltonian for the surface states was given by

$$\mathcal{H}_S = (\boldsymbol{\sigma} \times \boldsymbol{\kappa})_z \,. \tag{3.1.60}$$

The spectrum is a Dirac cone, $\varepsilon_\kappa = \pm\kappa$ and the corresponding spin textures are perpendicular to $\boldsymbol{\kappa}$ and go in opposite directions in the upper and lower cones (see the previous chapter)

$$\langle \boldsymbol{\sigma} \rangle = \pm\,(\sin\theta_{\boldsymbol{\kappa}}, -\cos\theta_{\boldsymbol{\kappa}}, 0)\,, \tag{3.1.61}$$

where the positive (negative) sign corresponds to the upper (lower) cone. Since momentum and spin are interrelated and are orthogonal, we said that this system exhibits spin-momentum locking. Moreover, we argued that the opposite signs for $\langle \boldsymbol{\sigma} \rangle$ corresponded to opposite helicities [19]. Let us now consider an interface where the gap changes from λ to $-\lambda$. In this case, following the same procedure discussed in the previous chapter, we obtain the following surface Hamiltonian

$$\mathcal{H}'_S = -\,(\boldsymbol{\sigma} \times \boldsymbol{\kappa})_z \,. \tag{3.1.62}$$

As we can observe, the two surface Hamiltonians are related by $\boldsymbol{\kappa} \to -\boldsymbol{\kappa}$. Hence, the spin texture also changes sign, since $\theta_{\boldsymbol{\kappa}} \to \theta_{\boldsymbol{\kappa}} + \pi$. Therefore, the helicities are inverted with respect to the case where we change from $-\lambda$ to λ. If we now consider a film, the gap will change from $-\lambda$ to λ and then from λ to $-\lambda$, so we will have that each surface is characterized by topological surface states of opposite helicities, as schematically depicted in Fig. 3.9. This change in sign in the helicity allows for annihilation of the two Dirac cones, leading to the massive doubly degenerate spectrum that we discussed above. In other words, states from different helicities mix and a gap opens up.

In this scenario, it seems that it would be interesting to try to minimize the gap. For that matter, we can apply the mechanism that we considered for the topological boundary. Namely, we can apply an electric field to achieve such a result. Indeed, say we have a boundary at $\xi = -a$ that changes from λ to $-\lambda$ and another boundary at $\xi = a$ that changes from $-\lambda$ to λ. Then, when applying the electric field potential, $f\xi$, the cones of each isolated boundary will widen (not considering the overlap yet), thereby reducing the Fermi velocity. Since the decay length is about v_F/λ, then by decreasing v_F we also decrease the decay length and, in consequence, the overlap between the two states. This results in a reduction of the energy gap. Hence, the electric field allows us to reduce this gap while maintaining the same width of the film. We may say that it behaves as an effective thickness. Let us explore this situation within the approximations of the single boundary. In this case, the only change to the differential equation resides on the perturbation term $U(\xi)$

$$U(\xi) = 2i\beta\alpha_z\,[\delta(\xi - a) - \delta(\xi + a)]\,. \tag{3.1.63}$$

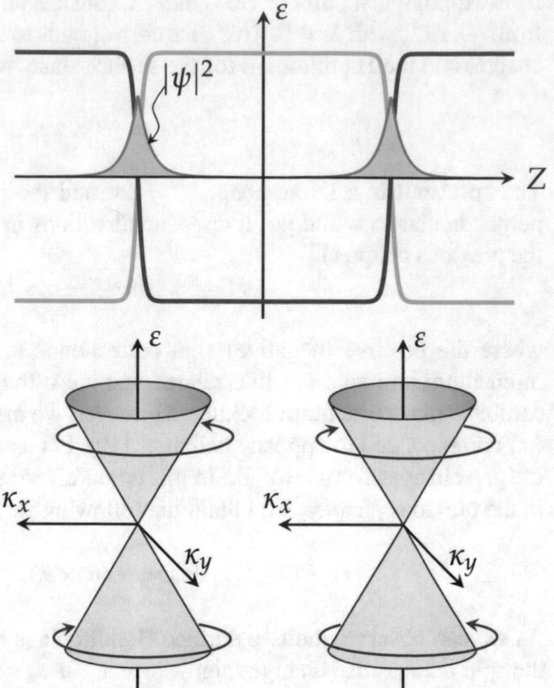

Fig. 3.9 Two topological boundaries hosting surface states of opposite helicities (curved arrows in opposite directions for the two cones). When brought close together, the spectrum becomes doubly degenerate and massive. See main text for details

Using the Lippman–Schwinger equation we find

$$\chi(z) = -2\mathrm{i}\,\beta\alpha_z\,[G_0(\xi, a; \varepsilon)\chi(a) - G_0(\xi, -a; \varepsilon)\chi(-a)]\,, \tag{3.1.64}$$

where we have already assumed G_0 to be a scalar function as in the previous section. Let us denote by $G_{\pm\pm} = G_0(\pm a, \pm a)$. Then, if we particularize the previous equation to $\xi = \pm a$, we find that

$$(\mathbb{1}_4 + 2\mathrm{i}\,G_{++}\beta\alpha_z)\,\chi(a) = 2\mathrm{i}\,G_{+-}\beta\alpha_z\chi(-a), \tag{3.1.65a}$$

$$(\mathbb{1}_4 - 2\mathrm{i}\,G_{--}\beta\alpha_z)\,\chi(-a) = -2\mathrm{i}\,G_{-+}\beta\alpha_z\chi(a). \tag{3.1.65b}$$

The asymptotic cases of $a \to 0$ and $a \to \infty$ in the field-free cases are readily understood from these two equations. In the first case, $\rho \simeq 0$ and $\chi(a) = \chi(-a)$, which implies that $\chi(z) \simeq 0$ in that case. That is, when the two boundaries of opposite chiralities coincide, the surface states are fully anihilated. The other situation has $\rho \simeq 1$ and decouples the two equations above, meaning that the two surfaces are uncoupled and the dispersion is that of Dirac cones. Intermediate situations can be explored by operating a little to obtain the following eigenvalue problem

$$\chi(a) = \mathcal{N}\chi(a), \tag{3.1.66}$$

where

$$\mathcal{N} = -\frac{1}{4G_{+-}G_{-+}}\left[\mathbb{1}_4\left(1 - 4G_{++}G_{--}\right) - 2\mathrm{i}\,\alpha_z\beta\left(G_{++} - G_{--}\right)\right]. \tag{3.1.67}$$

We must then retain only those eigenvalues of $\mathcal{N}$ that are equal to one. At this point, we can already see that in the field-free problem where $G_{++} = G_{--}$ the matrix is proportional to the identity. Hence, its four eigenvalues are fourfold degenerate and we must keep all four of them. This is in contrast to the topological boundary, where we only kept two eigenvalues. With this in mind, if we take into account the form of $G_{\pm\pm}$, we can then write straightforwardly that

$$\varepsilon^2 = \kappa^2 + \exp\left(-4\rho a\right). \tag{3.1.68}$$

Notice that we recover the asymptotic behaviours mentioned previously: the Dirac cone solutions $\varepsilon = \pm\kappa$ only holds if $a \to \infty$, that is, if the two boundaries do not see each other, and the surface states melt within the bulk when $a \to 0$. For an intermediate situation of a film that is not too narrow, we can approximate $\rho a \simeq a$ in the exponential and write

$$\varepsilon = \pm\sqrt{\kappa^2 + \exp\left(-4a\right)}. \tag{3.1.69}$$

As we can observe, the dispersion is now that of a massive Dirac fermion with a width-dependent mass, as expected,

$$\lambda_0 = \exp\left(-2a\right). \tag{3.1.70}$$

We can connect this result to what we said earlier about the helicities being coupled by using the surface effective Hamiltonians. Indeed, the Hamiltonian for the two helicities would in that case be given by [18]

$$\mathcal{H} = \tau_z \otimes \left(\boldsymbol{\sigma} \times \boldsymbol{k}\right)_z + m\tau_x \otimes \sigma_0, \tag{3.1.71}$$

where the τ matrices act on the helicity subspace and m simulates the coupling between the two surface states. Notice that such a coupling is compliant with time-reversal symmetry

$$\Theta = \tau_0 \otimes \mathrm{i}\,\sigma_y \mathcal{K}, \tag{3.1.72}$$

which takes $\boldsymbol{k} \to -\boldsymbol{k}$, and rotational symmetry about the Z-axis

$$\mathcal{R}(\theta) = \tau_0 \otimes \exp\left(\mathrm{i}\,\frac{\theta}{2}\sigma_z\right), \tag{3.1.73}$$

which takes $k_\pm \to \exp(\mp\mathrm{i}\,\theta)k_\pm$, as we saw in the previous chapter. The spectrum is easily obtained by squaring the Hamiltonian,

$$\varepsilon = \pm\sqrt{\kappa^2 + m^2}, \tag{3.1.74}$$

where each of the two massive Dirac bands are doubly degenerate. Since the strength of m depends on the overlap between the surface states and these are exponentially localized, it is plausible to argue that the hibridization gap, $2m$, should decay as well $m \simeq \exp(-2a)$. This way, the two approaches coincide. In order to tackle the problem of the electric field, we need the eigenvalues of $\mathcal{N}$, which in this case is no longer diagonal since $G_{++} \neq G_{--}$ in general. Such a matrix has two doubly degenerate eigenvalues and the requirement that they be equal to 1 implies

$$\pm 2\left[G_{++} - G_{--}\right] = 1 - 4G_{+-}G_{-+} - 4G_{++}G_{--}. \tag{3.1.75}$$

It must be noted that the Dyson equation leads to the same result [2]. Taking advantage of the fact that we know from the single boundary that the level widths are negligible, we will approximate $\mathrm{Ci}^{\pm}(x) \simeq \mathrm{Bi}\,(x)$, which will only render the real part of the energy. If we take into account the asymptotic expansions for the Airy functions introduced in the previous section, after some tedious algebra we arrive to

$$\varepsilon^2 \simeq \kappa^2 + \exp\left(-4\rho a\right) \mp \frac{2|\varepsilon| f a}{\rho}. \tag{3.1.76}$$

For not too narrow films, we may approximate $\rho \simeq 1$ and the equation can be solved to give

$$\varepsilon^{g}_{\kappa,s} \simeq s\left[\sqrt{(fa)^2 + \kappa^2 + \lambda_0^2} - fa\right], \tag{3.1.77}$$

and

$$\varepsilon^{e}_{\kappa,s} \simeq s\left[\sqrt{(fa)^2 + \kappa^2 + \lambda_0^2} + fa\right], \tag{3.1.78}$$

with $s = \pm 1$. Notice that the electric field is breaking the double degeneracy of the Dirac massive spectrum. Indeed, if $f = 0$, then we obtain the two doubly degenerate bands, since $\varepsilon^{g}_{\pm} = \varepsilon^{e}_{\pm}$ in that case. We show in Fig. 3.10 the two bands, as obtained by numerically solving Eq. (3.1.75), for two values of f and a. Additionally, we show in that same plot the corresponding approximate bands, as given by Eqs. (3.1.77) and (3.1.78). As it can be observed in panel (a), corresponding to $a = 0.5$, the agreement for low fields is significant, particularly for the lower band. However, it is even more noteworthy in panel (b), corresponding to $a = 1$. For low fields, the agreement holds perfectly for both bands, while the upper band is not correctly captured for high fields by the approximation.

In any case, we observe two main features. First, as we just said, the electric field leads to a splitting of the bands, which increases linearly with the field. This is shown in Fig. 3.11a, where the numerical result given in Eq. (3.1.75) is compared with the approximate level splitting as obtained from

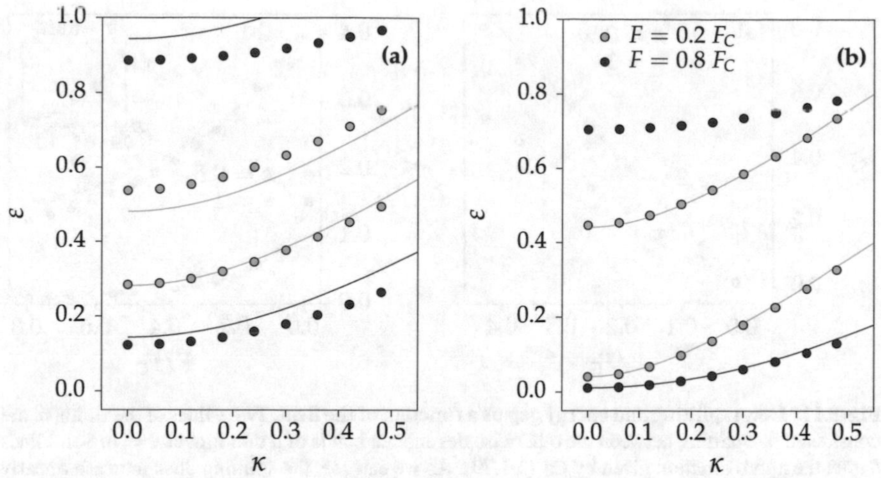

Fig. 3.10 Dispersion in a thin film with a perpendicular electric field applied. Two widths are considered, **a** $a = 0.5$ and **b** $a = 1$, for two values of the field. Solid lines depict the approximation given by Eqs. (3.1.77) and (3.1.78). As we can see in both panels, there are two bands that split apart with a splitting that increases with the field. Instead, the energy gap decreases with the field. The agreement is noteworthy for low fields and not too narrow widths

$$\delta_f = 2fa. \tag{3.1.79}$$

As we can observe, the agreement is noteworthy for $a = 1$.

The second main observation and probably the most important of the two is the behaviour of the energy gap. Indeed, it decreases with a, as it should do, but more interestingly it decreases with the field, as discussed earlier in this section. In fact, the approximated result (3.1.77) at $\kappa = 0$ predicts that the gap $2\lambda_f$ decreases with the field in the following fashion

$$\lambda_f = \sqrt{(fa)^2 + \lambda_0^2} - fa. \tag{3.1.80}$$

We can further distinguish two regimes: $fa < \lambda_0$, where the gap decreases linearly with the field $\lambda_f \simeq \lambda_0 - fa$, and the regime where $fa > \lambda_0$, where the gap decreases with the inverse of the field, that is, $\lambda_f \simeq \lambda_0^2/2fa$. We show in Fig. 3.11b the gap obtained from the numerical solution of (3.1.75) and that obtained from (3.1.80). As we can see, the gap does indeed decrease with the field and is smaller the larger a is. It is interesting to observe that, in the case where $a = 0.5$, there is no agreement between the approximation and the exact result at $f = 0$, where one would imagine that both should agree. However, one must bear in mind that the approximations hold only for not too narrow films where $\rho \simeq 1$, in which case the energy gap of the unbiased system is given by $2\lambda_0$. If instead one considers Eq. (3.1.76) where no approximations concerning the width of the films have been made, the agreement at

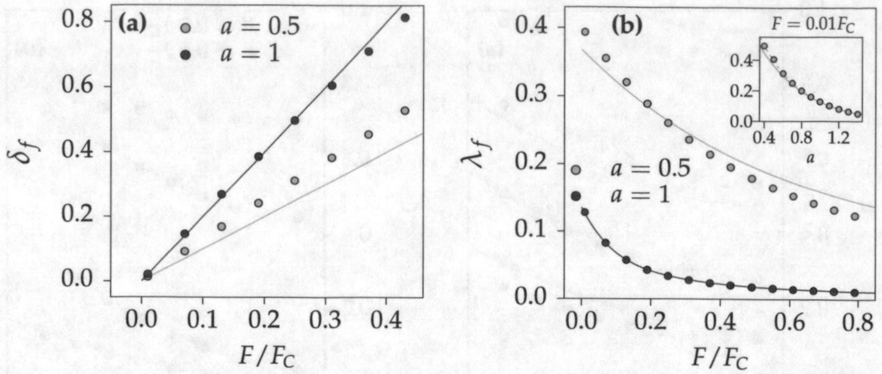

Fig. 3.11 **Level splitting and energy gap as a function of the field**. Two values of the width, a, are considered. **a** Splitting between the otherwise degenerate bands of a thin film at $\kappa = 0$. Solid lines depict the approximation given by Eq. (3.1.79). As we can see, the splitting does increase linearly with the field, the agreement with the approximated result being significant for $a = 1$. **b** Gap between the two massive Dirac bands closest to zero energy. Solid lines depict the approximation given by Eq. (3.1.80). It can be observed that the gap decreases with the field notably, the agreement with the approximated result being significant for $a = 1$. The inset shows the gap as a function of a for a very small field of $0.01F_C$ obtained by numerically solving (3.1.75). Solid line depicts the approximation given in (3.1.80) and dashed line depicts the solution to (3.1.76). It can be drawn that the latter coincides perfectly with the numerical solution at all distances and all three approaches coincide when the width is sufficiently large

$f = 0$ is restored. In fact, when $a \gtrsim 0.7$ the approximated result (3.1.77) coincides with the exact result (3.1.75) and the approximation that does not make assumptions on the width (3.1.76). This is shown in the inset of Fig. 3.11b.

3.1.4 Exact Solution

In the previous sections, we have made a number of approximations that, in turn, have enabled us to find simple analytic solutions to complete our understanding of the phenomenon. However, it would be interesting to see if the exact solution of the problem predicts the same behaviour. As we shall see, this solution does not allow us to obtain analytic expressions for the energies, but it does indeed predict the same physics as the ones presented in the topological boundary section. Moreover, we shall consider arbitrary-sized gaps. The equation we must then solve is

$$\left\{-\mathrm{i}\alpha_z\partial_\xi + \boldsymbol{\alpha}_\perp\cdot\boldsymbol{\kappa} + [\delta + s(\xi)]\beta + \gamma s(\xi) - \left[\varepsilon - \gamma\delta + f\xi\right]\right\}\chi(\xi) = 0, \tag{3.1.81}$$

where $s(\xi) = \mathrm{sgn}\,(\xi)$. Notice that we have changed $f \to -f$ with respect to the previous discussions. This should be taken into consideration when analysing the results of this section. The boundary conditions that we will apply will be: continuity at the interface, $\boldsymbol{\chi}(0^-) = \boldsymbol{\chi}(0^+)$, and the condition of vanishing of the current probability at some distance larger than the typical decay length of surface states, $L \gg 1$. This last condition is nothing but the infinite-mass condition that we talked about in the previous chapter in discussing the topological protection of Dirac cones in graphene armchair nanoribbons when using the continuum description. In this case, the infinite-mass condition is written very similarly to that case [20]

$$\boldsymbol{\chi}(\xi_b) = \mathcal{M}_{\infty}^{b} \boldsymbol{\chi}(\xi_b), \qquad \mathcal{M}_{\infty}^{b} = \tau_y \otimes (\boldsymbol{\sigma} \cdot \boldsymbol{n}_b), \tag{3.1.82}$$

where we are assuming a static boundary, b, with normal vector $\boldsymbol{n}_b$ pointing in the direction of the outgoing current. In our case, we have a boundary at $\xi_1 = -L$ and another at $\xi_2 = L$, so $\boldsymbol{n}_1 = -\hat{\boldsymbol{z}}$ and $\boldsymbol{n}_2 = \hat{\boldsymbol{z}}$. All in all, the boundary matrices read

$$\mathcal{M}_{\infty}^{1} = -\mathcal{M}_{\infty}^{2} = -\tau_y \otimes \sigma_z. \tag{3.1.83}$$

With this in mind, let us solve the problem at hand. In order to do so, it proves convenient to perform a rotation of π about the τ_z axis in the orbital subspace, so that $(\tau_x, \tau_y, \tau_z) \to (-\tau_x, -\tau_y, \tau_z)$, followed by a rotation of $\pi/2$ about the τ_y axis, so that $(-\tau_x, -\tau_y, \tau_z) \to (\tau_z, -\tau_y, \tau_z)$. All in all, both operations exchange τ_x and τ_z and change the sign of τ_y: $(\tau_x, \tau_y, \tau_z) \to (\tau_z, -\tau_y, \tau_x)$. The operator that allows us to achieve such a transformation would therefore be

$$\mathcal{U} = \exp\left(-\mathrm{i}\frac{\pi}{4}\tau_y\right) \tau_z \otimes \sigma_0. \tag{3.1.84}$$

Notice that we have used the fact that τ_z itself allows us to perform the operation of a rotation of π about the τ_z axis. This way, the $\boldsymbol{\alpha} \cdot \boldsymbol{\kappa}$ part of the Hamiltonian will become block diagonal. Although we will pay the price of not having β to be diagonal, it will prove to be useful to consider the transformed Hamiltonian. That is,

$$\mathcal{H}^{\mathcal{U}} = \tau_z \otimes \mathcal{H}_0 + \left[\gamma s(\xi) - f\xi\right] \tau_0 \otimes \sigma_0 + \left[\delta + s(\xi)\right] \tau_x \otimes \sigma_0, \tag{3.1.85}$$

where

$$\mathcal{H}_0 = -\mathrm{i}\,\sigma_z \partial_\xi + \boldsymbol{\sigma}_\perp \cdot \boldsymbol{\kappa}. \tag{3.1.86}$$

Hence, we have to solve

$$\left\{\tau_z \otimes \mathcal{H}_0 - \sqrt{f}x(\xi) + \left[\delta + s(\xi)\right] \tau_x \otimes \sigma_0\right\} \boldsymbol{\Phi}(\xi) = 0, \tag{3.1.87}$$

where $\boldsymbol{\Phi}(\xi) = \mathcal{U}^{-1}\boldsymbol{\chi}(\xi)$ and

$$x(\xi) = \frac{1}{\sqrt{f}}\left[\varepsilon - \gamma\delta - \gamma s(\xi) + f\xi\right]. \tag{3.1.88}$$

If we write $\boldsymbol{\Phi}$ as follows

$$\boldsymbol{\Phi} = \begin{pmatrix} \boldsymbol{\Phi}_u \\ \boldsymbol{\Phi}_l \end{pmatrix}, \tag{3.1.89}$$

then $\boldsymbol{\Phi}_u$ satisfies

$$\left[\partial_x^2 + x^2 - \mathrm{i}\,\sigma_z - 4\mu^2\right] \boldsymbol{\Phi}_u = 0, \tag{3.1.90}$$

where

$$\mu^2(\xi) = \frac{1}{4f}\left\{\kappa^2 + [\delta + s(\xi)]^2\right\}. \tag{3.1.91}$$

In order to arrive to Eq. (3.1.90), we have to take into account that we will solve on both sides of $\xi = 0$, where the sign function is constant. We can then obtain $\boldsymbol{\Phi}_l$ from $\boldsymbol{\Phi}_u$

$$\boldsymbol{\Phi}_l = -\frac{1}{\delta + s(\xi)}\left[-\sqrt{f}\,(\mathrm{i}\,\partial_x \sigma_z + x) + \boldsymbol{\sigma}_\perp \cdot \boldsymbol{\kappa}\right] \boldsymbol{\Phi}_u, \tag{3.1.92}$$

where x is given in Eq. (3.1.88).

Before we move on, it is interesting to notice that $\boldsymbol{\kappa}$ is coupling each of the two components of $\boldsymbol{\Phi}_l$ with the other two components of $\boldsymbol{\Phi}_u$. That is, if $\boldsymbol{\kappa} = 0$, then each component of $\boldsymbol{\Phi}_l$ is related to a single component of $\boldsymbol{\Phi}_u$. This is a manifestation of Kramers' theorem. It must be observed that the two independent solutions for the upper component of $\boldsymbol{\Phi}_u$ correspond to the same solutions but complex conjugated for the lower component of $\boldsymbol{\Phi}_u$. Similarly, one can obtain the lower component of $\boldsymbol{\Phi}_l$ by taking complex conjugates of the upper component of $\boldsymbol{\Phi}_l$ and taking care of the integration constants. It is then immediately shown that

$$\boldsymbol{\Phi}_u = \left(P, \sigma_x P^*\right) \cdot \boldsymbol{C}, \tag{3.1.93}$$

where $\boldsymbol{C}$ is a constant vector of four components and $P(x)$ is a matrix defined as

$$P(x) = \begin{pmatrix} F^*(x) & G(x) \\ 0 & 0 \end{pmatrix}, \tag{3.1.94}$$

and

$$F(x) = M\left(-\mathrm{i}\,\mu^2, \frac{1}{2}, \mathrm{i}\,x^2\right) e^{-\mathrm{i}x^2/2}, \tag{3.1.95a}$$

$$G(x) = -2\mathrm{i}\,\mu x\, M\left(1 - \mathrm{i}\,\mu^2, \frac{3}{2}, \mathrm{i}\,x^2\right) e^{-\mathrm{i}x^2/2}, \tag{3.1.95b}$$

where $M(a, b, z)$ are Kummer's functions [11]. The functions $F(x)$ and $G(x)$ satisfy the useful relations

$$(\mathrm{i}\,\partial_x + x)\, F^*(x) = 2\mu G(x), \tag{3.1.96a}$$

$$(\mathrm{i}\,\partial_x + x)\, G(x) = 2\mu F(x). \tag{3.1.96b}$$

Using these relations and Eqs. (3.1.92) and (3.1.93), we find

$$\mathbf{\Phi}_l = \left(\tau P^* \sigma_x + \eta \sigma_x P \;\; \tau \sigma_x P \sigma_x + \eta^* P^*\right) \mathbf{C}, \tag{3.1.97}$$

where we have introduced

$$\tau = \frac{2\mu\sqrt{f}}{\delta + s(\xi)}, \qquad \eta = -\frac{\kappa_x + \mathrm{i}\,\kappa_y}{\delta + s(\xi)}. \tag{3.1.98}$$

Finally, $\mathbf{\Phi}$ can be expressed as

$$\mathbf{\Phi}(x) = \mathcal{F}(x)\mathbf{C}, \qquad \mathcal{F}(x) = \begin{pmatrix} P & \sigma_x P^* \\ \tau P^* \sigma_x + \eta \sigma_x P & \tau \sigma_x P \sigma_x + \eta^* P^* \end{pmatrix}. \tag{3.1.99}$$

We must now impose the boundary conditions at the interface and at $\xi = \pm L$. We have 4 constants encoded in $\mathbf{C}$, and there are two such vectors on both sides of the interface. Hence, there are 8 constants. Continuity at the interface leads to 4 equations. There are another 4 equations at L and at $-L$, leading to a total of 12 equations. However, there are redundancies in the equations at $\pm L$, which reduce the number of equations to 8 in total. Hence, we obtain a homogeneous system of equations, which has to be such that the determinant of the matrix of coefficients vanishes to discard trivial solutions. After some tedious algebra, we arrive at the following condition [3]

$$\det\left[\begin{pmatrix} \mathcal{F}(x_0^+) & -\mathcal{F}(x_0^-) \\ \mathcal{P}^+(x_L^+) & (\tau_x \otimes \sigma_0)\mathcal{P}^-(x_L^-) \end{pmatrix}\right] = 0, \tag{3.1.100}$$

with $x_0^\pm = x(\xi = 0^\pm)$, $x_L^\pm = x(\xi = \pm L)$ and

$$\mathcal{P}^\pm(x) = \begin{pmatrix} (\eta\sigma_x \pm \mathrm{i}\,\sigma_z)\, P + \tau P^* \sigma_x & \left(\eta^* \mp \sigma_y\right) P^* + \tau \sigma_x P \sigma_x \\ \mathbf{0}_2 & \mathbf{0}_2 \end{pmatrix}. \tag{3.1.101}$$

The method described in this section, although being the most general one, has its subtleties. On the one hand, in the absence of field, the energies of the states in the continuum cannot take values within the gap. However, when the field is turned on, the band edges tilt and these states can have energies that are within the gap, although being well separated from the surface states for small fields. On the other hand, placing the system in a box leads to a quantization of the continuum energies, which in turn leads to subbands. Therefore, in order to extract meaningful data from the surface states, one can observe which energies are unaffected when changing the size of the box. Indeed, the quantization of the subbands in the continuum results from quantum confinement in the Z-direction, but the surface states are already confined and have a rapid decay. Thus, if the system has a length $L \gg 1$ with L in units of d, one will observe a set of energies for different values of $\boldsymbol{\kappa}$ that do not change upon changing L, whereas the remaining energies do change. The latter corresponds

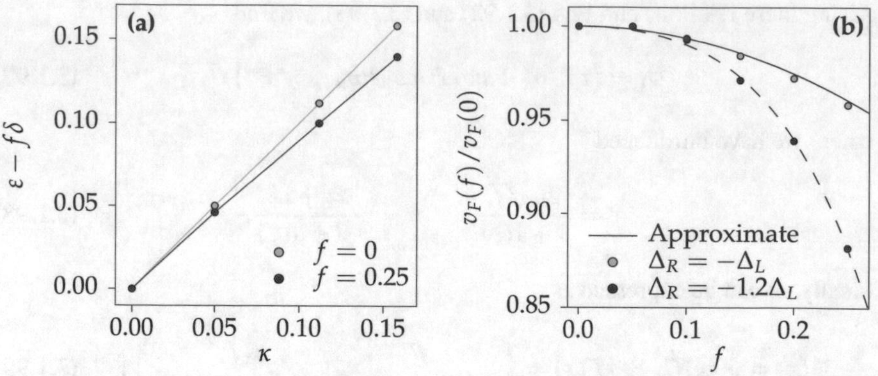

Fig. 3.12 Dispersion and Fermi velocity as a function of the external field. a Dispersion for an asymmetric-centered scenario with $\Delta_R = -1.2\Delta_L$ for both zero and nonzero fields. We have substracted the correction obtained from first order perturbation theory, $f\delta$. Solid lines are a guide to the eye. **b** Fermi velocity as a function of the field for centered-symmetric ($\Delta_R = -\Delta_L$) and asymmetric ($\Delta_R = -1.2\Delta_L$ in this case) setups. Solid line corresponds to the approximated result (3.1.59) and dashed line is a fit in even powers of f up to f^4

to the subbands, the former to the surface states. Hence, for small fields and large values of L, one will observe perfect crossings between the cone and a large amount of energies coming from the bulk. Since there are no energy gaps at these crossings, one can affirm that the surface state remains well localized and does not hybridize with states from the subbands. This, of course, only holds for small fields. Larger fields lead to larger band-edge tiltings and, as a result, there will be hybridization. In this case, extracting meaningful data is far more complicated, since all states get mixed. Anyhow, it is the low field limit that is interesting as it is the experimentally feasible.

With this result, we can now try to observe if the Fermi velocity is also reduced in a centered junction with different-sized gaps, which was not captured by perturbation theory. In Fig. 3.12a, we show the dispersion for two values of the field when the gaps differ by a 20%, that is, $\Delta_R = -1.2\Delta_L$. Two main features can be observed. First, the unbiased cone widens and moves downwards in energy when the field is applied. This fact we knew from perturbation theory and it is accounted for by removing $f\delta$ from the energy. This way, it is easier to observe the second feature, that is, that the cone does indeed widen, leading to a reduction in the Fermi velocity. In Fig. 3.12b, we show the reduction of the velocity in the centered symmetric and asymmetric situations. First of all, it must be noted the perfect agreement with the approximated result (3.1.59) in the case of a symmetric junction, which is shown as a solid line. Second, the velocity decreases with even powers of f up to f^4 in the asymmetric junction, as shown by the fitted dashed line. As was already apparent by our discussion of the effect in the approximate solution for the centered-symmetric system, one would have to extend perturbation theory to second order with terms proportional to f^2 to capture the effect of a Fermi velocity reduction.

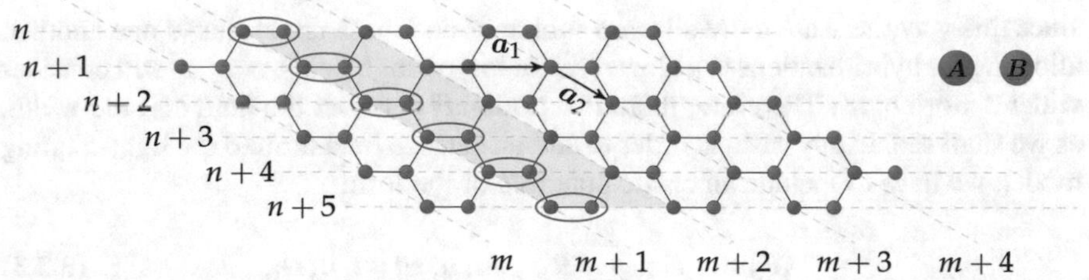

Fig. 3.13 Armchair nanoribbon. The horizontal direction is taken to be periodic and the vertical direction is finite. The ovals show a way to arrange the A and B atoms. A supercell is shown in yellow. The lattice vectors are $\boldsymbol{a}_1$ and $\boldsymbol{a}_2$

3.2 Metallic Armchair Graphene Nanoribbons

In this section, we will consider the effect of an electric field across the transverse direction of metallic armchair graphene nanoribbons. A quick reminder from the second chapter is in order. For that matter, we reproduce the lattice shown in the second chapter in Fig. 3.13.

The lattice is infinite along the longitudinal direction and finite along the transverse direction. A supercell shown in yellow contains N_y pairs of A-B atoms. The lattice constant is $\sqrt{3}$ times larger than that of bulk graphene, which is given by the distance between next-nearest-neighbours in the honeycomb lattice. The lattice can be generated by the following two vectors in units of the lattice constant

$$\boldsymbol{a}_1 = (1, 0), \qquad \boldsymbol{a}_2 = \frac{1}{2\sqrt{3}}\left(\sqrt{3}, -1\right), \tag{3.2.1}$$

so that a position in the lattice is given by

$$\boldsymbol{R}_{m,n} = m\boldsymbol{a}_1 + n\boldsymbol{a}_2, \qquad m \in \mathbb{Z}, \qquad n = 1, \ldots, N_y. \tag{3.2.2}$$

The A and B atoms within an oval are linked by $\boldsymbol{\delta} = (1/3, 0)$, such that A atoms are at $\boldsymbol{R}^A = \boldsymbol{R}_{m,n}$ and B atoms at $\boldsymbol{R}^B = \boldsymbol{R}_{m,n} + \boldsymbol{\delta}$. The wavevectors k live within the first Brillouin zone, $k \in [-\pi, \pi)$. As we know from the second chapter, a first-nearest neighbours model renders a spectrum of subbands due to momentum quantization resulting from quantum confinement along the vertical direction. In fact, the eigenstates have the shape of standing waves. Quantization along the transverse direction may or may not result in cutting or not through the Dirac point. The former case leads to metallic nanoribbons and is achieved whenever $N_y = 3r - 1$ with $r = 1, 2, \ldots$. It is also important to remember that, in contrast to the zigzag and bearded nanoribbons, there is no topological protection of the armchair nanoribbon within the tight-binding description. Therefore, one would expect that upon applying an electric field would open up a gap in the otherwise gapless Dirac-like spectrum of the lowest subband. That happens indeed, the effect being more pronounced the wider the nanoribbon is, since this way there are more allowed subbands and these are closer to one another, allowing for hybridizations [1]. However, for low enough fields, one can still consider wider nanoribbons. Therefore, there is a trade-off between the field and the width, as we shall see. In any case, in order to add an electric field term to the tight-binding model, we have to include an onsite potential of the form

$$V(F) = e\boldsymbol{F} \cdot \sum_{m,n,\alpha} \boldsymbol{R}^{\alpha}_{m,n} |m, n; \alpha\rangle \langle m, n; \alpha|. \tag{3.2.3}$$

Notice that, as written, eF is a dimensionless quantity, since we are considering $a = 1$ and the potential is written in units of $t = 1$. Since $\boldsymbol{F}$ points along the transverse direction, there is still translational symmetry along the longitudinal direction, so we can still make use of the good quantum number k along that direction. Moreover, atoms within an oval feel the same potential since they are at the same transverse position. In fact, the nontrivial part of $V(F)$ will be simply given by

$$V(F) = -\frac{eF}{2\sqrt{3}} \sum_{n} n|n\rangle \langle n|. \tag{3.2.4}$$

This is what we would expect by taking into account Fig. 3.13. Indeed, if we recall that we are expressing distances in units of the lattice constant and that this is in turn $\sqrt{3}$ times larger than the bulk lattice constant, a, this term is telling us that upon moving from one row to the next there is a potential drop of $eFa/2$. Since we will consider lattices with $N_y = 3r - 1$, which is an odd number, it is convenient to set the potential to zero right at the middle row, $n = (N_y + 1)/2$. Alternatively, one can redefine n to start at $-(N_y + 1)/2$ and end at $(N_y + 1)/2$. Doing so, the tight-binding model can be written as follows

$$\frac{eF}{2\sqrt{3}} n\, \psi_A(n) + e^{\mathrm{i}k/2} \psi_B(n) + \psi_B(n-1) + \psi_B(n+1) = E\psi_A(n), \tag{3.2.5a}$$

$$\frac{eF}{2\sqrt{3}} n\, \psi_B(n) + e^{-\mathrm{i}k/2} \psi_A(n) + \psi_A(n+1) + \psi_A(n-1) = E\psi_B(n). \tag{3.2.5b}$$

It is important to remember that energies are expressed in units of the hopping, $t = 1$, so that the Fermi velocity of the unbiased nanoribbon is $v_F = 1/2$. In order to compare different combinations for the fields and widths, it is convenient to introduce a few more quantities. Let $W = (N_y - 1)/(2\sqrt{3})$ be the width of the nanoribbon. Since we nullify the wavefunction at fictitious sites at $n = 0$ and $n = N_y + 1$, the width of the fictitious nanoribbon is $\widetilde{W} = (N_y + 1)/(2\sqrt{3})$. Then, it proves to be convenient to write

$$f = \frac{F}{F_W}, \qquad F_W = \frac{1}{2e\widetilde{W}^2} = \frac{6}{e(N_y + 1)^2}. \tag{3.2.6}$$

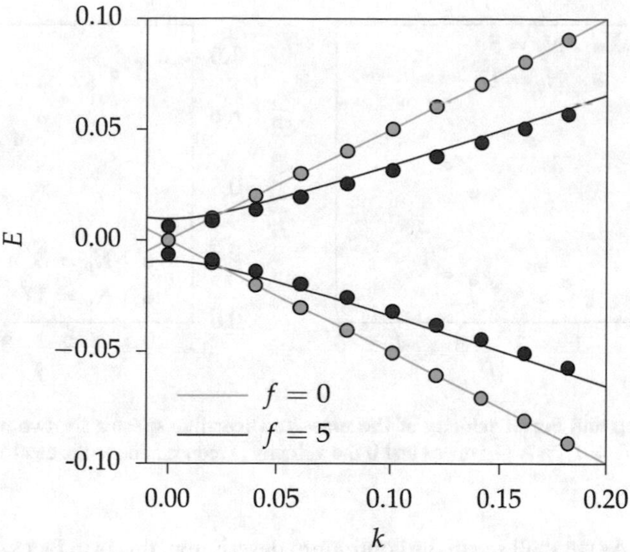

Fig. 3.14 Dispersion of the lowest subbands of a nanoribbon with $N_y = 11$ for $f = 0$ and $f = 5$. Solid lines correspond to nonlinear fits to massive Dirac-like spectra. Notice the gap opening at $k = 0$ and the reduction of the Fermi velocity

This shows the trade-off that we were talking about previously: one can have large values of N_y and small values of F and, similarly, have small values of N_y for large values of F in order to have the same f. In other words, in order to have two nanoribbons of different width with the same potential difference between the two sides of the ribbon, the field has to change accordingly, as we would expect. With this in mind, let us solve the tight-binding model numerically for a representative nanoribbon of, say, $N_y = 11$, for two values of f. The results are shown in Fig. 3.14, where the data are fit to massive Dirac dispersions

$$E^2 = (\alpha k)^2 + m^2. \tag{3.2.7}$$

Notice that α is the Fermi velocity and $\Delta = 2m$ is the energy gap (in units of t). In unbiased graphene, $\alpha = 1/2$ and $m = 0$. However, in biased graphene, $m \neq 0$ and the nanoribbon becomes semiconducting.

From the fit, we can plot both the gap and the Fermi velocity as a function of f for two values of N_y. As we can observe in Fig. 3.15a, the gap increases with f. It is important not to draw wrong conclusions from this figure by forgetting that f is a quantity that depends on N_y and it is not the electric field itself. That is, even though in the plot the gap increases more slowly for $N_y = 17$ than $N_y = 5$, also the real fields are smaller in the $N_y = 17$ case. The velocity reduction is also nicely observed in Fig. 3.15b. Interestingly enough, there is not much dependence on N_y but rather

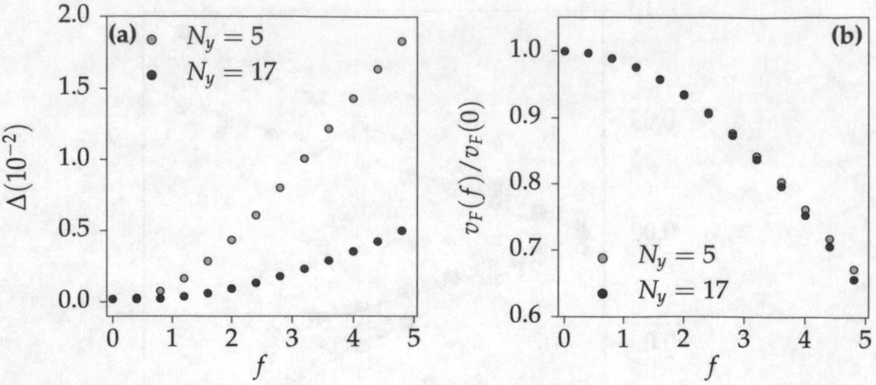

Fig. 3.15 Gap and Fermi velocity of the massive Dirac-like spectra for two nanoribbons of $N_y = 5$ and $N_y = 17$. **a** A gap opens and **b** the velocity is reduced due to the electric field

on f alone. As we shall see in the continuum description, this is in fact why we chose to define f as we did.

Finally, before we move on to the continuum description, it is worth trying to understand the reduction of the Fermi velocity in this setup. For that matter, we shall consider the effect of the field on the eigenstates for the positive and negative energies of a representative k. In particular, we shall consider $N_y = 17$, $f = 5$ and $k = 0.3$. The states are displayed in Fig. 3.16. It is important to mention that the horizontal axis does not correspond to the row. Rather, if n is the row index, then $m = 2n - 1$ is the position of the position of the A atoms and $m = 2n$ is the position of the B atoms. Notice that the probability density of A and B atoms of the same row is the same, as it should from the symmetry of the nanoribbon.

As it can be observed in the figure, the two eigenstates of positive and negative energy of the unbiased nanoribbon (dashed lines) have the same probability amplitudes. However, once the field is turned on, states of higher energies move towards the left-hand side, where the electric potential is more negative, whereas states of lower energies do the exact opposite. This is very much like what we discussed in the previous section, where states try to redistribute their probability distributions in order to minimize the effects of the electric field. As a result of moving towards regions of lower potential, states of higher energies move down in energy, and the opposite happens for states of lower energies. The effect is most notable for states of larger momenta and, as in the topological boundary, momentum is conserved and the net result is a reduction in energy of all states in such a way that we still get a Dirac-like dispersion with a reduced Fermi velocity. In contrast to the topological boundary, a gap opens up due to the lack of topological protection. We shall see shortly that the continuum description, which is topologically protected as we discussed in the second chapter, does indeed protect the Dirac point, while reducing the Fermi velocity as well.

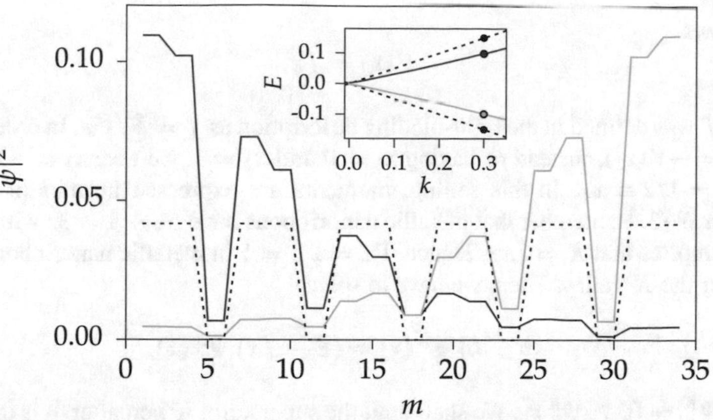

Fig. 3.16 Probability density for an unbiased (dashed lines, black) and a biased nanoribbon (coloured curves). Here $N_y = 17$ and $f = 5$. The inset shows the dispersion and dots are the states shown on the figure. Positive energies of the biased nanoribbon are dark-red coloured and negative energies are blue coloured

Let us then explore the continuum model. For that matter, we will write the Hamiltonian in the basis $\{\psi_A^K(x), \psi_B^K(x), -\psi_B^{K'}(x), \psi_A^{K'}(x)\}$, which leads to the isotropic Hamiltonian for both valleys

$$\mathcal{H} = \tau_0 \otimes (\boldsymbol{\sigma}_\perp \cdot \boldsymbol{k}) + \tau_0 \otimes \sigma_0 V(x), \tag{3.2.8}$$

where we have taken $v_F(0) = 1$ to be the Fermi velocity in the absence of field. Here, in contrast to the tight-binding model, X is the transverse direction instead of Y. The physics will obviously be the same, just like in the previous chapter. We will however write N_x instead of N_y and the results from the previous section are recovered by writing $N_y \to N_x$. This is of course not necessary, but we will do it to stick to the conventions chosen in the previous chapter. As we know, the boundary conditions do not require to nullify the envelope functions, but rather the coefficients in the expansion of atomic orbitals (see previous chapter for details). The resulting boundary condition is then

$$\psi_\alpha^K(x_b) = -\Gamma_b \psi_\alpha^{K'}(x_b), \qquad \Gamma_b = \exp(-\mathrm{i}2Kx_b). \tag{3.2.9}$$

where $x_1 = 0$ and $x_2 = (N_x + 1)/2$ are the boundary fictitious sites where nullification takes place. Notice that we are expressing distances in units of the bulk lattice constant, $a = 1$. Here, $K = 4\pi/3$. Taking into account that we are working with the isotropic Hamiltonian, we can solve for either of the two valleys and obtain the other one by simply changing the integration constants. Instead of continuing to work in units of a, which in the continuum version does not make a lot of sense anyway, it is convenient to express distances in units of $\tilde{W}$. This way, the potential $V(x)$ is written as follows

$$V(x) = fx, \tag{3.2.10}$$

where f was defined in the tight-binding description as $f = F/F_W$. In order to have $V(x_1) = -V(x_2)$, instead of having $x_1 = 0$ and $x_2 = 1$, we take $x_1 = -1/2 \equiv x_-$ and $x_2 = 1/2 \equiv x_+$. In this setting, momenta are expressed in units of $1/\tilde{W}$, so $K = 4\pi \tilde{W}/3$. Remember that metallic nanoribbons have $N_y + 1 = 3r$ with $r \in \mathbb{Z}^+$, which implies that $K = 2\pi r$. Hence, $\Gamma_+ = \Gamma_- = 1$ in metallic nanoribbons. Let us consider the K valley. Then, we have to solve

$$(\boldsymbol{\sigma}_\perp \cdot \boldsymbol{k})\, \boldsymbol{\Psi}^K(x) = (E - fx)\, \boldsymbol{\Psi}^K(x). \tag{3.2.11}$$

where $\boldsymbol{\Psi}^K = (\psi_A^K, \psi_B^K)^T$. We shall omit the superscript K hereafter. It is interesting to perform a rotation in the sublattice space that takes σ_x to σ_z. We do so by rotating $\pi/2$ around the σ_y axis with the operator

$$\mathcal{R} = \exp\left(-\mathrm{i}\,\frac{\pi}{4}\sigma_y\right). \tag{3.2.12}$$

Hence, we have

$$\left(-\mathrm{i}\,\partial_x \sigma_z + k\sigma_y\right) \boldsymbol{\chi}(x) = (E - fx)\, \boldsymbol{\chi}(x), \tag{3.2.13}$$

where $\boldsymbol{\chi}(x) = \mathcal{R}^{-1}\boldsymbol{\Psi}(x)$. Notice that this rotation does not mix the valley indices since it is diagonal in that subspace. Therefore, we may still identify $\boldsymbol{\chi}^K$ and $\boldsymbol{\chi}^{K'}$. Let

$$z = \frac{1}{\sqrt{f}}(E - fx), \qquad \mu = \frac{k}{2\sqrt{f}}. \tag{3.2.14}$$

Then, we can write Eq. (3.2.13) as follows

$$\left(-\mathrm{i}\,\partial_z \sigma_z + 2\mu\sigma_y\right) \boldsymbol{\chi}(z) = z\boldsymbol{\chi}(z). \tag{3.2.15}$$

If we act on the left with $(-\mathrm{i}\,\partial_z \sigma_z + 2\mu\sigma_y)$, we obtain

$$\left[\partial_z^2 + z^2 - 4\mu^2 - \mathrm{i}\right] \boldsymbol{\chi}(z) = 0. \tag{3.2.16}$$

As in the previous chapter, it may seem that the two components of $\boldsymbol{\chi}$ are decoupled and satisfy the same equation. However, the two are actually coupled by

$$\chi_l = -\frac{1}{2\mathrm{i}\,\mu}(\mathrm{i}\,\partial_z + z)\,\chi_u, \tag{3.2.17}$$

where $\boldsymbol{\chi} = (\chi_u, \chi_l)^T$. Hence, we have to solve

$$\left[\partial_z^2 + z^2 - 4\mu^2 - \mathrm{i}\right] \chi_u(z) = 0, \tag{3.2.18}$$

to obtain later χ_l from (3.2.17). This equation is already familiar to us from the exact solution of the topological boundary. Indeed, two independent solutions are given in terms of Kummer's functions $M(a, b, z)$ [11]

$$F(z) = M\left(-\mathrm{i}\,\mu^2, \frac{1}{2}, \mathrm{i}\,z^2\right) e^{-\mathrm{i}\,z^2/2}, \tag{3.2.19a}$$

$$G(z) = 2\mathrm{i}\,\mu z M\left(1 - \mathrm{i}\,\mu^2, \frac{3}{2}, \mathrm{i}\,z^2\right) e^{-\mathrm{i}\,z^2/2}. \tag{3.2.19b}$$

Hence, $\chi_u(z)$ is given by

$$\chi_u(z) = \alpha F^*(z) + \beta G(z). \tag{3.2.20}$$

In order to obtain χ_l, we can make use of the following useful relations

$$(\mathrm{i}\,\partial_z + z)\, F^*(z) = -2\mu G^*(z), \quad (\mathrm{i}\,\partial_z + z)\, G(z) = -2\mu F(z). \tag{3.2.21}$$

Thus, it is straightforward to obtain

$$\chi_l(z) = -\mathrm{i}\,\alpha G^*(z) - \mathrm{i}\,\beta F(z). \tag{3.2.22}$$

Solutions for K' are obtained from these solutions by changing the integration constants. Hence, we may write

$$\boldsymbol{\chi}^{K(K')}(z) = \mathcal{P}(z)\boldsymbol{C}_{K(K')}, \qquad \mathcal{P}(z) = \begin{pmatrix} F^*(z) & G(z) \\ -\mathrm{i}\,G^*(z) & -\mathrm{i}\,F(z) \end{pmatrix}, \tag{3.2.23}$$

where $\boldsymbol{C}_{K(K')}$ is a constant vector of two components containing the integration constants, $\alpha_{K(K')}$ and $\beta_{K(K')}$. We now have to apply the boundary conditions. Notice that a rotation in the sublattice space does not alter the boundary conditions. Indeed, originally we had for the boundary conditions

$$\boldsymbol{\Psi}^K(x_\pm) = -\mathrm{i}\,\Gamma_\pm \sigma_y \boldsymbol{\Psi}^{K'}(x_\pm), \qquad \Gamma_\pm = \exp\left[\mp\mathrm{i}\,\frac{2\pi}{3}(N_x + 1)\right], \tag{3.2.24}$$

where σ_y appears due to the ordering of the basis. Since the rotation is performed around the σ_y axis, then σ_y is not affected by the rotation and we may write directly

$$\boldsymbol{\chi}^K(z_\pm) = -\mathrm{i}\,\Gamma_\pm \sigma_y \boldsymbol{\chi}^{K'}(z_\pm), \tag{3.2.25}$$

where $z_\pm = z(x_\pm)$. These two equations (one for each sign) lead to the following equation for $\boldsymbol{C}_K$

$$\Gamma_-^2 \boldsymbol{C}_K = \mathcal{T}\boldsymbol{C}_K, \tag{3.2.26}$$

being

$$\mathcal{T} = \mathcal{P}^{-1}(z_+)\sigma_y\mathcal{P}(z_+)\mathcal{P}^{-1}(z_-)\sigma_y\mathcal{P}(z_-). \tag{3.2.27}$$

Here we have used the fact that $\Gamma_+^{-1} = \Gamma_-$. Notice that we have assumed the existence of the inverse of $\mathcal{P}$. However, this is always the case since we can prove the determinant of $\mathcal{P}$ to be a constant for all z. That is, we want to prove

$$\partial_z \left[\det \mathcal{P}(z)\right] = 0. \tag{3.2.28}$$

Taking into account the relations (3.2.21), it is straightforward to prove that

$$\partial_z |F(z)|^2 = \partial_z |G(z)|^2 = 2\mathrm{i}\,\mu \left[F(z)G^*(z) - F^*G(z)\right]. \tag{3.2.29}$$

Since

$$\det \mathcal{P}(z) = -\mathrm{i}\left(|F(z)|^2 - |G(z)|^2\right), \tag{3.2.30}$$

it is then clear that $\det \mathcal{P}(z)$ is a constant. That is, the determinant is independent of z, so we can calculate it for any z of our choice. Particularly simple is the case where $z = 0$ since $M(a, b, 0) = 1$, which in turn leads to

$$\det \mathcal{P}(z) = -\mathrm{i}\,. \tag{3.2.31}$$

Therefore, we conclude that it is always possible to define the inverse of $\mathcal{P}$. Equation (3.2.26) requires

$$\det\left[\Gamma_-^2 \mathbb{1}_2 - \mathcal{T}\right] = 0, \tag{3.2.32}$$

Equivalently,

$$\Gamma_-^4 - \mathrm{Tr}(\mathcal{T})\Gamma_-^2 + \det \mathcal{T} = 0. \tag{3.2.33}$$

Taking into account the definition of $\mathcal{T}$, we find that $\det \mathcal{T} = 1$ and, therefore,

$$\left(\Gamma_-^2 + \frac{1}{\Gamma_-^2}\right) = \mathrm{Tr}(\mathcal{T}). \tag{3.2.34}$$

Using the definition of Γ_-, we find

$$\cos\left[\frac{4\pi}{3}(N_x + 1)\right] = \frac{1}{2}\mathrm{Tr}(\mathcal{T}). \tag{3.2.35}$$

It is interesting to make a few observations of this equation. If $N_x = 3r - 1$, with $r \in \mathbb{Z}^+$, that is, when the nanoribbon is metallic, then we have

$$\frac{1}{2}\mathrm{Tr}(\mathcal{T}) = 1. \tag{3.2.36}$$

In any other case, we find

$$\mathrm{Tr}(\mathcal{T}) = -1. \tag{3.2.37}$$

Since these equations are independent of N_x, it seems that the spectrum of subbands will be the same for all values of N_x for all metallic and semiconducting nanoribbons. However, this is not the case since momenta are expressed in units of $1/\tilde{W}$ and do depend on N_x. In any case, these two equations show the usefulness of working with dimensionless variables, since it allows us to consider different sizes all at once. It is not difficult to find that the trace of $\mathcal{T}$ is given by

$$\mathrm{Tr}\,(\mathcal{T}) = 2\Re\left[\left(F_+^2 - G_+^2\right)\left((F_-^*)^2 - (G_-^*)^2\right)\right] - 8\Im\left[F_+G_+^*\right]\Im\left[F_-G_-^*\right], \tag{3.2.38}$$

where $F_\pm = F(z_\pm)$ and $G_\pm = G(z_\pm)$. Before we solve this equation numerically, it is interesting to consider what happens at $k = 0$. Indeed, we argued in the second chapter that the Dirac point should remain protected in the continuum model even in presence of a potential $V(x)$. If $k = 0$, then it is straightforward to show that

$$\mathrm{Tr}(\mathcal{T}) = 2\cos\left(z_+^2 - z_-^2\right). \tag{3.2.39}$$

Since $z_+^2 - z_-^2 = -2E$ and it is independent of f, the energy should be the same at all fields. Even though we cannot consider the case $f = 0$ since there is a singularity at that point, the result holding for any f means that it holds for an f that is infinitely close to $f = 0$, and by adiabatic continuity it must happen that the energy at $f = 0$ and at $f \to 0$ coincide. Indeed, in metallic nanoribbons we find

$$E = n\pi, \qquad n \in \mathbb{Z}, \tag{3.2.40}$$

so the Dirac point $n = 0$ remains a solution, as expected. Moreover, it can be shown numerically that degenerate subbands above and below the Dirac non-degenerate bands remain degenerate upon application of the field. That is, the perturbation does not lift the pseudovalley degeneracy introduced in the second chapter. Indeed, we can prove this generally by solving the problem numerically, but it is analytically accesible in the $k = 0$ case. Metallic nanoribbons satisfy

$$1 - \cos(z_+^2 - z_-^2) = 0. \tag{3.2.41}$$

Since the term on the left is bounded to be within $[0, 2]$, then $E = n\pi$ are double roots and there is double degeneracy. In contrast, semiconducting nanoribbons satisfy

$$1 + 2\cos(z_+^2 - z_-^2) = 0. \tag{3.2.42}$$

Here, the term on the left is bounded to be within $[-1, 3]$, so the roots are single roots and there is no degeneracy. Numerically, one can show that the subbands of metallic nanoribbons remain degenerate away from $k = 0$, except for the Dirac band which is non-degenerate, save for the Dirac point at $k = 0$ as we have just shown.

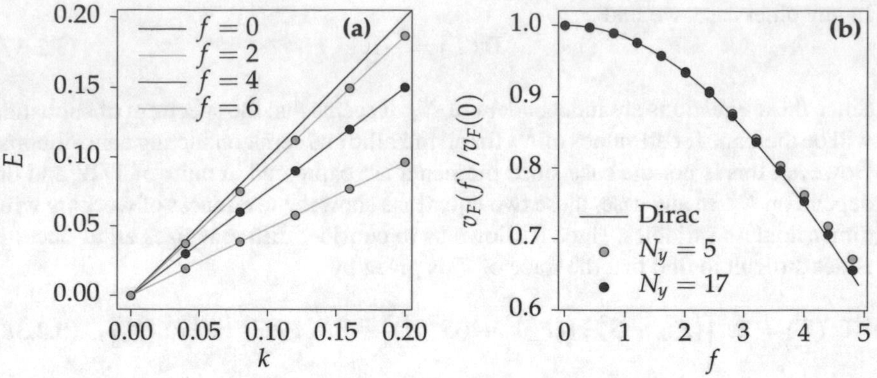

Fig. 3.17 Dirac band of metallic armchair nanoribbons and Fermi velocity reduction. **a** As the electric field increases, the Dirac dispersion tilts while preserving the Dirac point. **b** Fermi velocity reduction as a function of the field as obtained from tight-binding calculations (dots) for two values of N_y and a low-energy Dirac-like Hamiltonian (solid line)

Finally, in Fig. 3.17a, we show the expected behaviour of preserving the Dirac point while reducing the slope as the field is increased in metallic nanoribbons. What is even more interesting is to compare the reduction of the Fermi velocity as predicted by both the tight-binding and continuum approaches. This is shown in Fig. 3.17b, where the agreement is noteworthy between both approaches.

3.3 Conclusions

In this chapter, the most prominent result is the ability to modify the Fermi velocity of two paradigmatic Dirac systems, namely, a topological insulator and a metallic armchair nanoribbon. The results shown in this chapter seem to indicate that this behaviour is generic to other Dirac materials (see Ref. [21] for a detailed review on Dirac matter). The Fermi velocity reduction has been exploited to reduce the energy gap that occurs due to the hybridization of surface states in a thin film, ultimately linked to the annihilation of opposite helicities. Although it has not been explicitly mentioned throughout the text, the phenomena presented herein can be collectively gathered in a class of *quantum confined Stark effects* [22, 23]. This effects are most well-known from its presence in biased quantum wells, where upon applying an electric field the electron levels experience a reduction of their energy that goes with F^2 at low field amplitude, F being the field amplitude. The phenomenon occurs due to the fact that the wavefunction moves to regions of lower potential, very much like what we have observed in graphene. Moreover, bound states in the quantum well can tunnel into the continuum since there is a finite barrier that separates these two types of states. As a result, these bound states are no longer bound states, but rather *resonant states*, very much like those discussed in the topological boundary. It is interesting to observe that perturbation theory did not provide us any information about tunneling into the continuum, also known as Fowler-Nordheim tunneling [23]. The presence of tunneling is also absent when doing perturbation theory in the most simple setup of an ordinary quantum well [23]. This is telling us that perturbation theory cannot be applied to capture the full physics, and thus one has to solve the problem in full, as we did in the section where we discussed an approximate solution. This problem is not encountered in the case of graphene nanorribbons since the states have no continuum states to resonate with. Finally, before we conclude this chapter, it is worth mentioning that the Fermi velocity reduction has also been observed to occur in metallic carbon nanotubes [1]. It has also been confirmed in armchair graphene nanoribbons by means of density functional theory, although it is not possible to quantitatively compare with the results presented herein due to the presence of electronic correlations, electric polarizability and charge screening effects [1]. In the next chapter, we shall combine the ideas of applying an electric field by introducing a magnetic field as well.

References

1. Díaz-Fernández A, Chico L, González JW, Domínguez- Adame F (2017) Tuning the Fermi velocity in Dirac materials with an electric field. Sci Rep 7:8058
2. Díaz-Fernández A, Chico L, Domínguez-Adame F (2017) Electric control of the bandgap in quantum wells with band-inverted junctions. J Phys: Condens Matter 29:475301
3. Díaz-Fernández A, Domínguez-Adame F (2017) Quantum confined Stark effect in band-inverted junctions. Physica E: Low Dimens Syst Nanostruc 93:230
4. Black J, Conwell EM, Seigle L, Spencer CW (1957) Electrical and optical properties of some MVI..B 2 NVI..B 3 semiconductors. J Phys Chem Solids 2:240
5. Tchoumakov S, Jouffrey V, Inhofer A, Plaçis B, Carpentier D, Goerbig MO (2017) Volkov-Pankratov states in topological heterojunctions. Phys Rev B 96:201302
6. Korenman V, Drew HD (1987) Subbands in the gap in inverted-band semiconductor quantum wells. Phys Rev B 35:6446
7. Ludviksson A (1987) A simple model of a decaying quantum mechanical state. J Phys A: Math Gen 20:4733
8. Emmanouilidou A, Reichl LE (2000) Scattering properties of an open quantum system. Phys Rev A 62:022709
9. Jung J-W, Na K, Reichl LE (2009) Decay properties and photodetachment of the diatomic oxygen ion O_2 in a constant electric field. Phys Rev A 80:012518
10. Economou EN (2006) Green's functions in quantum physics. Springer, Berlin
11. Abramowitz M, Stegun I (1972) Handbook of mathematical functions. Dover, New York
12. Zhou B, Lu H-Z, Chu R-L, Shen S-Q, Niu Q (2008) Finite size effects on helical edge states in a quantum spin-hall system. Phys Rev Lett 101:246807
13. Linder J, Yokoyama T, Sudbo A (2009) Anomalous finite size effects on surface states in the topological insulator Bi_2Se_3. Phys Rev B 80:205401
14. Liu C-X, Zhang H, Yan B, Qi X-L, Frauenheim T, Dai X, Fang Z, Zhang S-C (2010) Oscillatory crossover from two-dimensional to threedimensional topological insulators. Phys Rev B 81:041307
15. Lu H-Z, Shan W-Y, Yao W, Niu Q, Shen S-Q (2010) Massive Dirac fermions and spin physics in an ultrathin film of topological insulator. Phys Rev B 81:115407
16. Shan W-Y, Lu H-Z, Shen S-Q (2010) Effective continuous model for surface states and thin films of three-dimensional topological insulators. New J Phys 12:043048

17. Buczko R, Cywiński Ł (2012) PbTe/PbSnTe heterostructures as analogs of topological insulators. Phys Rev B 85:205319
18. Ortmann F, Roche S, Valenzuela SO (2015) Topological insulators. Wiley, New Work
19. Zhang F, Kane CL, Mele EJ (2012) Surface states of topological insulators. Phys Rev B 86:081303
20. Alberto P, Fiolhais C, Gil MSV (1996) Relativistic particle in a box. Eur J Phys 17:19
21. Wehling TO, Black-Schaffer AM, Balatsky AV (2014) Dirac materials. Adv Phys 63:1
22. Bastard G (1991) Wave mechanics applied to semiconductor heterostructures. Les Editions de Physique, Les Ulis
23. Davies JH (1997) The physics of low-dimensional semiconductors: an introduction. Cambridge University Press, Cambridge

Chapter 4
Reshaping of Dirac Cones by Magnetic Fields

The physics of electrons in a magnetic field led to the discovery of quantum Hall phases in the 1980s and has led to three Nobel Prizes in Physics:

- **1985**: Klaus von Klitzing, for the experimental discovery of the integer quantum Hall effect [1].
- **1998**: Daniel C. Tsui and Horst L. Störmer, for the experimental discovery of the fractional quantum Hall effect [2] and Robert B. Laughlin, for theoretical investigations on this effect [3].
- **2016**: David J. Thouless, F. Duncan M. Haldane and J. Michael Kosterlitz for theoretical discoveries on topological quantum matter. Among other phenomena, the first two explored the physics of the quantum Hall effect and made seminal contributions to it [4–7].

These discoveries boosted the field of topology in condensed matter up to this day. In this chapter, we shall study the physics of topological surface states in a magnetic field, focusing on the effect on the Dirac cones. Moreover, we shall introduce an additional electric field perpendicular to the surface, in the spirit of the previous chapter. Before doing so, we provide a brief introduction to the theory of Landau levels.

If we think of electrons classically as rotating in a plane perpendicular to a uniform magnetic field, we may describe such a rotation as two perpendicular harmonic oscillators. Translated to the quantum world, kinetic energy gets quantized, leading to a stair-like spectrum of so-called Landau levels

$$E_n = \omega_c \left(n + \frac{1}{2} \right) , \qquad \omega_c = \frac{eB}{m^*} , \tag{4.0.1}$$

where B is the magnetic field strength and m^* is the effective mass. ω_c is referred to as the cyclotron frequency and coincides with the frequency of rotation in the classical setup if electrons are given a mass m^*. Notice that we are considering electrons confined to two dimensions or, alternatively, we are focusing on the lowest subband

Á. Díaz Fernández, *Reshaping of Dirac Cones in Topological Insulators and Graphene*, Springer Theses, https://doi.org/10.1007/978-3-030-61555-0_4

of a quantum well in the direction parallel to the field. The fact that the energies depend on a single quantum number n is inconsistent with the fact that there are two degrees of freedom. However, this is not the case if one takes into account that each Landau level turns out to be macroscopically degenerate. As usual, this degeneracy is a consequence of a symmetry. This symmetry cannot be the usual translational symmetry, where translation operators form an Abelian group and, therefore, upon performing a closed loop the wavefunction returns back to itself. Indeed, we know that a particle in the background of a gauge potential acquires an Aharonov-Bohm phase, so the translation operators would have to account for such a phase. This is achieved by replacing the two ordinary translation operators by another two operators such that they commute with the Hamiltonian, but not among themselves, so as to produce the Aharonov-Bohm phase. These *magnetic translation operators* would therefore form a non-Abelian group. This implies that we can find a common basis of eigenstates for the Hamiltonian and only one of the two translation operators, but not both at the same time. As a result, each of those eigenstates will be labeled by the eigenvalue of the Hamiltonian, n, and the eigenvalue of the magnetic translation operator of our choice, l. Therefore, a particular energy E_n is associated to a set of eigenstates, each of which is labeled by a distinct l. The total degeneracy can be found by studying how many distinct values of l there are [8]. In any case, however, it is simpler to provide with an intuitive explanation of the degeneracy. Let us imagine the harmonic oscillator. As we know, there is a characteristic length, $\ell = (m\omega)^{-1/2}$, such that the position uncertainty is $\Delta X \simeq \ell$. In the case of a two dimensional electron gas in a magnetic field, m is the effective mass m^*, ω is the cyclotron frequency, ω_c, and ℓ is known as the *magnetic length*,

$$\ell_B = \sqrt{\frac{1}{eB}} \,. \tag{4.0.2}$$

This is the characteristic length scale for all quantum effects where magnetic fields are involved. As it should, it diverges as the field goes to zero (in such a limit, momentum is again a good quantum number). If the field is given in tesla, it is convenient to write $\ell_B = 26\,\text{nm}/\sqrt{B}$. This implies that fields of the order of a tesla lead to nanometer-sized magnetic lengths, where quantum effects are most prominent. If we imagine placing the cyclotron orbit centered at a position (X, Y), we may say that such a position is encircled by an area of uncertainty of radius ℓ_B. Doing the proper calculations, it turns out that the radius is not ℓ_B, but rather $\sqrt{2}\ell_B$. The total degeneracy can then be obtained by dividing the total area into sectors of area $2\pi\ell_B^2$

$$D = \frac{A}{2\pi\ell_B^2} \,. \tag{4.0.3}$$

Since A is macroscopic (it is the sample's area) and ℓ_B is nanoscopic, this degeneracy is actually very large. There is another way as to interpret this degeneracy, which allows for the introduction of the Dirac flux quantum that appeared in the second chapter. Indeed, one can interpret the area $2\pi\ell_B^2$ as the area which, if threaded by a magnetic field, leads to the smallest possible flux, the flux quantum. That is,

$$\Phi_0 = 2\pi \ell_B^2 B = \frac{2\pi}{e} . \tag{4.0.4}$$

Hence, the degeneracy is counting the total number of flux quanta threading the sample. Notice as well that if B becomes too large, then ℓ_B may reach the angstrom scale, meaning that the underlying lattice can no longer be accounted for by simply performing the substitution $m \to m^*$. This implies the usage of extremely large magnetic fields. Alternatively, one can create superlattices where there is an effective lattice spacing that is much larger than the real lattice spacing [9–11], or by creating optical lattices [12–15] where the lattice spacing can be made much larger. In any case, solving a simple one-band tight-binding model including a magnetic field, as the flux increases the band splits forming a structure that resembles a Cantor set for each value of the flux. Taking into account that the system is invariant under $\Phi \to p\Phi_0$, where Φ is the flux through a unit cell and p is an integer, since in such a case the Aharonov-Bohm phases around the unit cell equal the identity, we can think of Φ as being periodic in Φ_0. If the spectrum is plotted as a function of $\Phi \in [0, \Phi_0)$, the result is a fractal-like structure named after its discoverer: the Hofstadter butterfly [16]. An example of the Hofstadter butterfly for a square lattice is shown in Fig. 4.1. For small values of the flux, the spectrum of Landau levels is recovered, as it can be seen in the lower left corner of the figure, where the linear dependence with the field is observed.

For some specific values of the flux, one can enlarge the unit cell so that it contains an integer multiple of flux quanta. In such a case, the Aharonov-Bohm phases around such unit cells equals the identity and one can find translation operators that commute among themselves and with the Hamiltonian, allowing to obtain energy bands as in the field-free case. That is, one can imagine the lattice undergoing some kind of renormalization upon where the new lattice has unit cells enclosing an integer number of flux quanta. As a result, one can find two momenta to label the energies as in the field-free problem and obtain a spectrum of energy bands. If the number of unit cells gathered in the enlarged unit cell is q, then there must be q energy bands. Indeed, the fact that the unit cell is now q times larger implies that the Brillouin zone is q times smaller, leading to the so-called *magnetic Brillouin zone*. Therefore, each band can accommodate N/q states, where N is the total number of unit cells of the original lattice. In order to accommodate the total N states of the original band, there must be q bands. As an example, when $\Phi = \Phi_0/2$, one needs an enlarged unit cell containing two original unit cells, so there should be two bands. Indeed, this is what we observe in Fig. 4.1.

Let us now turn our attention to the case of relativistic fermions or, more appropriately in our settings, massless quasiparticles that satisfy the Dirac equation. In this case, it is also possible to find a structure of Landau levels. However, the energy scale here can no longer be ω_c since $m^* = 0$. Indeed, it has to be replaced by v_F/ℓ_B. More precisely [17],

$$\omega_c = \sqrt{2}\, \frac{v_F}{\ell_B} . \tag{4.0.5}$$

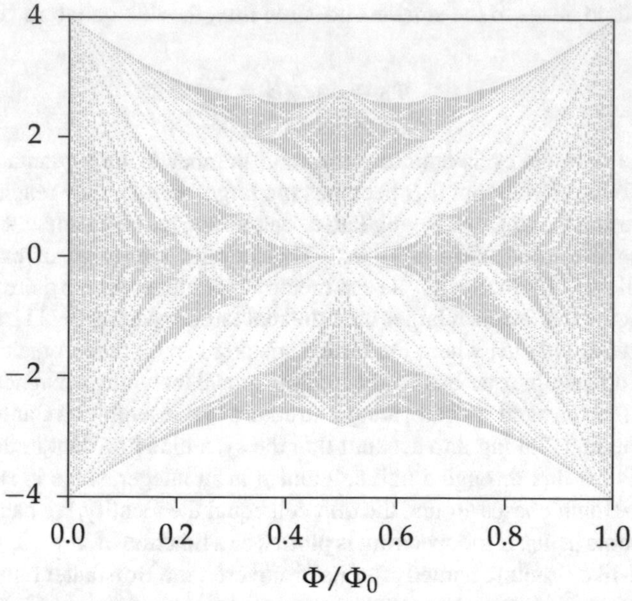

Fig. 4.1 Hofstadter butterfly in a square lattice. For zero flux, the cosine band of the square lattice ranging from −4 to 4 (hopping $t = 1$) is recovered. For low values of the flux at the lower left corner, the spectrum of Landau levels dispersing linearly with magnetic field in the continuum approximation is also recovered. The clearer regions are bulk energy gaps and the energies within them correspond to edge states (the problem has been solved on a cylinder)

Moreover, since the energy goes with k instead of k^2, we would expect the dispersion to go with $\sqrt{n}$. This is indeed what happens. Let us study this for the topological surface states by considering the effective Hamiltonian

$$\mathcal{H}_S = v_F \left(\boldsymbol{\sigma} \times \boldsymbol{k}\right)_z . \tag{4.0.6}$$

As long as ℓ_B is larger than the lattice spacing, we can still consider this continuum approach. In order to include a magnetic field, one has to write the previous equation in terms of the mechanical momentum

$$\boldsymbol{\Pi} = \boldsymbol{k} + e\boldsymbol{A} . \tag{4.0.7}$$

By doing so, the Hamiltonian has the same form as it would in the absence of a magnetic field

$$\mathcal{H}_S^B = v_F \left(\boldsymbol{\sigma} \times \boldsymbol{\Pi}\right)_z . \tag{4.0.8}$$

Taking into account the definition of $\boldsymbol{\Pi}$, it is straightforward to show that its components do not commute

$$[\Pi_x, \Pi_y] = -\mathrm{i}\,\frac{1}{\ell_B^2}\,. \tag{4.0.9}$$

The approach now is the same that one would follow with ordinary Landau levels: introduce a pair of raising and lowering operators, $a^\dagger$ and a, such that the mechanical momenta are written as follows

$$\Pi_x = \frac{1}{\sqrt{2}\ell_B}\left(a^\dagger + a\right)\,, \qquad \Pi_y = \frac{1}{\mathrm{i}\sqrt{2}\ell_B}\left(a^\dagger - a\right). \tag{4.0.10}$$

This way, the newly introduced operators satisfy

$$\left[a, a^\dagger\right] = 1\,. \tag{4.0.11}$$

Expressing the Hamiltonian in terms of these operators we find

$$\mathcal{H}_S^B = \mathrm{i}\,\omega_c \begin{pmatrix} 0 & a \\ -a^\dagger & 0 \end{pmatrix}, \tag{4.0.12}$$

where ω_c is defined by Eq. (4.0.5). The effective Hamiltonian acts upon a spinor ψ

$$\mathcal{H}_S^B \psi = E\psi\,. \tag{4.0.13}$$

If we square the Hamiltonian, we obtain the following result

$$\omega_c^2 \begin{pmatrix} 1+N & 0 \\ 0 & N \end{pmatrix} \psi = E^2 \psi\,, \tag{4.0.14}$$

where $N = a^\dagger a$ is the number operator. It is then clear that if we choose

$$\psi = \alpha \begin{pmatrix} \phi_{n-1} \\ 0 \end{pmatrix} + \beta \begin{pmatrix} 0 \\ \phi_n \end{pmatrix}, \tag{4.0.15}$$

with ϕ_n an eigenstate of the number operator, $N\phi_n = n\phi_n$ with $n \geq 0$, then the energy is found to be $E^2 = (\omega_c n)^2$, that is,

$$E_{n,s} = s\omega_c\sqrt{n}\,, \tag{4.0.16}$$

with $s = \pm 1$. It is interesting to observe the presence of a zero energy Landau level, which is totally absent from the ordinary Landau levels of a semiconductor and which is independent of the magnetic field strength. In fact, this Landau level is special in yet another sense: it has support only on the lower component of ψ, since $\phi_n = 0$ for $n < 0$. It can also be shown [18] that the degeneracy of each Landau level is still given by D [cf. Eq. (4.0.3)]. Hence, the Dirac point turns from being a point of zero density of states to having a Landau level with a large degeneracy. This

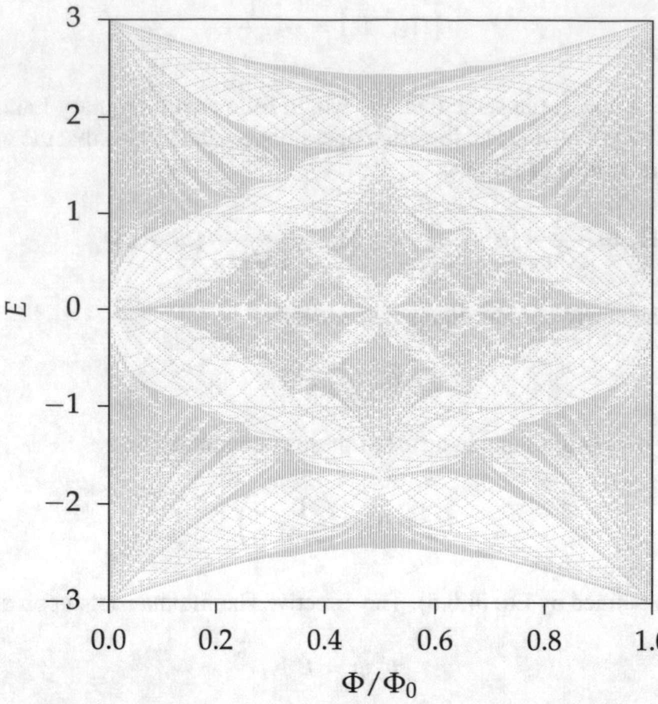

Fig. 4.2 Hofstadter butterfly in a honeycomb lattice. For zero flux, the π and π^* bands of the honeycomb lattice ranging from -3 to 3 (hopping $t = 1$) are recovered. For low values of the flux and energies around $E = 0$, we observe the squared-root-like dependence with the field, as well as the field-independent zero Landau level of the continuum approach. The clearer regions are bulk energy gaps and the energies within them correspond to edge states (the problem has been solved on a cylinder)

feature is common to other Dirac materials, most notably in graphene, where the main consequence is the so-called *anomalous integer quantum Hall effect*. Although we shall not go into the details, the presence of this Landau level leads to a large peak in the longitudinal conductance at charge neutrality, which occurs right at the Dirac point in graphene. This is because the Landau level contributes with extended states that are available for transport. Hence, at the same time, right at charge neutrality there is a jump in the transverse conductivity to reach a plateau when a mobility gap is reached, thereby disregarding the possibility for a plateau at zero conductance, in contrast to ordinary semiconductors. We shall not go into the details of the quantum Hall effect in graphene and refer the reader to Refs. [17, 18]. This effect has also been experimentally detected in Bi_2Se_3 [19, 20]. It is also interesting to compare the continuum approximation with the tight-binding description in the case of graphene. In doing so, we obtain another fractal-like structure [21], as shown in Fig. 4.2.

Coming back to our discussion, if we take into account the definition of ω_c, we can write the energy as follows

$$E_{n,s} = s v_F \sqrt{2eBn}\,. \tag{4.0.17}$$

This result is already interesting for our discussion. Indeed, if we consider the results of the second chapter, one may argue that since the effect of the electric field is to renormalize the Fermi velocity (at least to low fields and momenta), while preserving the cones, then it should be possible to use the effective Hamiltonian that we have used by changing $v_F \to v_F(f)$. Hence, we would expect that Landau level spectroscopy as discussed in Refs. [19, 20] should unravel the veracity of our results. In fact, in Refs. [22, 23], the authors discuss the case of graphene under a perpendicular magnetic field and an in-plane electric field. As they argue in their paper using arguments from special relativity, the presence of an electric field leads to the same set of Landau levels of graphene in the absence of electric field, although with a reduced effective magnetic field. This is precisely what is happening if we do $v_F \to v_F(f)$, since we can write

$$E^F_{n,s} = s v_F(f)\sqrt{2eBn} = s v_F \sqrt{2eB(f)n}\,, \tag{4.0.18}$$

where

$$B(f) = B\left(1 - \frac{5}{8}f^2\right)^2\,. \tag{4.0.19}$$

Therefore, the magnetic field would be effectively reduced by the electric field, in agreement to Refs. [22, 23].

4.1 General Formalism

In this chapter, we shall consider the setup to be a centered-symmetric topological boundary upon which an electric field perpendicular to the boundary is applied, as in the previous section. Additionally, we will include in-plane and out-of-plane magnetic fields. By performing the substitution of momentum to mechanical momentum, the Hamiltonian reads

$$\mathcal{H}_B = \boldsymbol{\alpha}\cdot\boldsymbol{\Pi} + \beta\,\mathrm{sgn}\,(z)\,, \tag{4.1.1}$$

where energies are measured in units of half the energy gap Δ, the mechanical momentum is measured in units of the inverse decay length of unperturbed states and $d = v_F/\Delta$, z is measured in units of d. Notice that, if $\boldsymbol{A}$ is the vector potential, the dimensionless version would be $\widetilde{\boldsymbol{A}} = (1/dB)\boldsymbol{A}$. Hence, the dimensionless mechanical momentum can be written as

$$\boldsymbol{\Pi} = \boldsymbol{p} + b\widetilde{\boldsymbol{A}}\,, \qquad b = \frac{d^2}{\ell_B^2}\,. \tag{4.1.2}$$

Notice that the dimensionless magnetic field depends on the ratio between the decay and magnetic lengths. Since d is of the order of a few nanometers, the continuum approximation makes sense as long as $b < 1$. Instead of following an algebraic approach as in the previous section, we will choose a specific gauge to work with. Taking into account the geometry of our problem, it makes sense to consider two different gauges: a symmetric and a Landau gauge. The former will be used when studying the case of a magnetic field perpendicular to the boundary, in order to exploit the cylindrical symmetry of the problem. The latter will be used when discussing an in-plane magnetic field. Let $\widehat{\boldsymbol{B}}$ be the direction of the magnetic field. Then, the Landau gauge reads

$$\widetilde{\boldsymbol{A}} = x_j \left(\widehat{\boldsymbol{B}} \times \widehat{\boldsymbol{x}}_j\right) , \tag{4.1.3}$$

where there is no summation over repeated indices. Here, $x_1 = x$, $x_2 = y$, $x_3 = z$, and $\widehat{\boldsymbol{x}}_j$ is a unit vector along the j-th direction. Notice that there are two choices for the Landau gauge, those where $\widehat{\boldsymbol{x}}_j$ and $\widehat{\boldsymbol{B}}$ are perpendicular. We will choose the one that simplifies our calculations. The symmetric gauge reads

$$\widetilde{\boldsymbol{A}} = \frac{1}{2}\left(\widehat{\boldsymbol{B}} \times \boldsymbol{r}\right) , \tag{4.1.4}$$

where $\boldsymbol{r} = (x, y, z)$.

The Hamiltonian acts on a bispinor $\boldsymbol{\Psi}(\boldsymbol{r})$ such that

$$\mathcal{H}_B \boldsymbol{\Psi}(\boldsymbol{r}) = (\varepsilon - f z)\, \boldsymbol{\Psi}(\boldsymbol{r}) , \tag{4.1.5}$$

where f is the electric field measured in units of $F_C = \Delta/ed$. Remember that we can interpret $f/2$ as being the ratio between the potential drop across the decay length d and the energy gap. As always, we square the Hamiltonian. We can reuse what we have found in the previous chapter, except that studying the square of $\boldsymbol{\alpha} \cdot \boldsymbol{\Pi}$ has to be done with a little care. In order to do so, it is interesting to take into account that

$$\alpha_i \alpha_j = \delta_{ij} \tau_0 \otimes \sigma_0 + \mathrm{i}\, \epsilon_{ijk} \tau_0 \otimes \sigma_k , \tag{4.1.6}$$

and

$$\widehat{B}_i = \epsilon_{ijk} \partial_j \widetilde{A}_k , \tag{4.1.7}$$

where ϵ_{ijk} is the Levi-Civita symbol, $\widehat{B}_i$ and $\widetilde{A}_i$ are the i-th components of $\widehat{\boldsymbol{B}}$ and $\widetilde{\boldsymbol{A}}$, respectively, and we are assuming sum over repeated indices. As a result, the squared Hamiltonian leads to

$$\begin{aligned}\Big[\big(\boldsymbol{p} + b\widetilde{\boldsymbol{A}}\big)^2 + b\tau_0 \otimes \big(\boldsymbol{\sigma} \cdot \widehat{\boldsymbol{B}}\big) + 1 - \varepsilon^2 + U(z) \\ - \mathrm{i}\, f \alpha_z + 2\varepsilon f z - f^2 z^2 \Big] \boldsymbol{\Psi}(\boldsymbol{r}) = 0 ,\end{aligned} \tag{4.1.8}$$

where

$$U(z) = 2\mathrm{i}\,\beta\alpha_z\delta(z)\,. \tag{4.1.9}$$

In the next sections, we will particularize this expression to the case of out-of-plane and in-plane magnetic fields. We shall refer to the first situation as parallel fields and the second situation as crossed fields.

4.2 Electron States Under Parallel Electric and Magnetic Fields

In this section, we consider a magnetic field along the Z-direction, in order to see if we obtain the result that we proposed in the introduction. That is, we aim to obtain Landau levels with an electric-field dependent Fermi velocity. In order to do so, we shall make use of the approximate solution that we presented in the previous chapter, so we will disregard the term $-\mathrm{i}\,f\alpha_z$ and treat $U(z)$ as a perturbation. This way, Eq. (4.1.8) becomes diagonal and we find four uncoupled equations for the components of $\boldsymbol{\Psi}$ which can be solved by separation of variables. The symmetry of the problem suggests the usage of cylindrical coordinates, where the vector potential in the symmetric gauge takes the form

$$\widetilde{\boldsymbol{A}} = \frac{1}{2}\rho\widehat{\boldsymbol{e}}_\phi\,. \tag{4.2.1}$$

In this gauge, $\boldsymbol{k}$ and $\boldsymbol{A}$ commute and we find that

$$\boldsymbol{p}\cdot\boldsymbol{A} + \boldsymbol{A}\cdot\boldsymbol{p} = -\mathrm{i}\,\partial_\phi \equiv L_z\,, \tag{4.2.2}$$

where we have identified the Z-component of angular momentum. Hence, Eq. (4.1.8) can be written as follows

$$\begin{aligned}\Big\{-\Big[\partial_\rho^2 + \rho^{-1}\partial_\rho - \rho^{-2}L_z^2\Big] + bL_z + \frac{b^2}{4}\rho^2 + b\tau_0\otimes\sigma_z \\ -\,\partial_z^2 + 1 - \varepsilon^2 - \mathrm{i}\,f\alpha_z + 2\varepsilon f z - f^2z^2\Big\}\boldsymbol{\Psi}(\boldsymbol{r}) = 0\,.\end{aligned} \tag{4.2.3}$$

Notice that the operator acting upon $\boldsymbol{\Psi}(\boldsymbol{r})$ commutes with L_z, that is L_z is a conserved quantity as a result of cylindrical symmetry. Hence, we can take $\boldsymbol{\Psi}(\boldsymbol{r})$ to be eigenstates of L_z with eigenvalue m, which allows us to write $\boldsymbol{\Psi}(\boldsymbol{r}) = \exp(\mathrm{i}\,m\phi)\chi(\rho, z)$. It is now convenient to expand $\boldsymbol{\Psi}(\boldsymbol{r})$ in the basis of eigenstates of $\tau_0\otimes\sigma_z$. We shall denote a generic eigenstate as φ_j, with eigenvalue $\lambda_j = (-1)^{j+1}$, where $j = 1, 2, 3$ and 4. Then,

$$\chi(\rho, z) = \sum_j h_j(\rho, z)\varphi_j\,. \tag{4.2.4}$$

As a result, we can write Eq. (4.2.3) as follows

$$\begin{aligned}\Big\{-\left[\partial_\rho^2+\rho^{-1}\partial_\rho+\rho^{-2}m^2\right]+\frac{b^2}{4}\rho^2+\lambda_j b\\ -\partial_z^2+1-\varepsilon^2-\mathrm{i}\,f\alpha_z+2\varepsilon f z-f^2z^2\Big\}h_j(\rho,z)=0\,.\end{aligned} \tag{4.2.5}$$

Notice that the upper line would correspond to the Hamiltonian of a two-dimensional harmonic oscillator in polar coordinates of mass m and frequency $\omega = b$, whereas the lower line corresponds to the problem of the perpendicular electric field described in the previous chapter. On the other hand, we see that $h_1(\rho, z)$ and $h_3(\rho, z)$ satisfy the same equation, and the same comment applies to the other two components. Moreover, the perturbation $U(\xi)$ only mixes the components i and $(i + 2)$ mod 4. Hence, we can safely choose the functions h_j to be eigenfunctions of the harmonic oscillator operator, such that the problem would now read

$$\left[-\partial_z^2+1-\varepsilon^2+\nu_j-\mathrm{i}\,f\alpha_z+2\varepsilon f z-f^2z^2\right]h_j(\rho,z)=0\,, \tag{4.2.6}$$

where

$$\nu_j=\lambda_j b+2b\left(n+\frac{1}{2}\right)\,,\qquad n\geq 0\,. \tag{4.2.7}$$

It is interesting to observe that it is only possible to have $\nu_j = 0$ if $j = 2$ or 4. This is what we would expect, since it would correspond to the zero Landau level in the absence of electric field and, as we saw in the introduction, it can only have support for the spin-down components. Hence, we can write

$$\nu=2bn\,, \tag{4.2.8}$$

with the knowledge that $\nu = 0$ can only have support for the spin-down components of the bispinor. With this result, the problem can therefore be considered to be the same as that with the electric field, although instead of κ^2 we have ν. Hence, we can take the results of the previous chapter for the dispersion and change κ by $\sqrt{\nu}$, that is,

$$\varepsilon^B_{n,s}=s\sqrt{2bn}\left(1-\frac{5}{8}f^2\right)\,. \tag{4.2.9}$$

Alternatively,

$$E^B_{n,s}=s v_F(f)\sqrt{2eBn}\,, \tag{4.2.10}$$

as predicted in the introduction of this chapter.

4.3 Electron States Under Crossed Electric and Magnetic Fields

In this section, we will apply an in-plane magnetic field, together with an electric field perpendicular to the boundary. We shall work in the Landau gauge, since it simplifies the calculations in this geometry. Let $\widehat{\boldsymbol{B}} = \widehat{\boldsymbol{y}}$. This choice respects translational symmetry in the XY-plane, but breaks rotational symmetry. As we said earlier, we have two main choices for the Landau gauge

$$\widetilde{\boldsymbol{A}}_x = z\left(\widehat{\boldsymbol{B}} \times \widehat{\boldsymbol{z}}\right) = z\,\widehat{\boldsymbol{x}}\,, \tag{4.3.1}$$

and

$$\widetilde{\boldsymbol{A}}_z = x\left(\widehat{\boldsymbol{B}} \times \widehat{\boldsymbol{x}}\right) = -x\,\widehat{\boldsymbol{z}}\,. \tag{4.3.2}$$

These two are, as all choices of $\widetilde{\boldsymbol{A}}$ are, connected by a gauge transformation. Indeed, we can write

$$\widetilde{\boldsymbol{A}}_x = \widetilde{\boldsymbol{A}}_z + \nabla g\,, \qquad g(x,z) = xz\,. \tag{4.3.3}$$

We shall choose $\widetilde{\boldsymbol{A}}_x$, so that we retain translational symmetry in the XY-plane, parallel to the topological boundary. This way, κ_x and κ_y will still be good quantum numbers. Notice, however, that rotational symmetry is broken, and we can no longer expect the dispersion to depend only on powers of κ. It is interesting to observe that the vector potential can only couple to the directions perpendicular to the magnetic field, thereby leaving the direction parallel to the magnetic field unaltered. This behaviour is the usual in magnetic fields, even classically, where the Lorentz force in the direction of the field is zero. As a result, we expect that, in absence of electric field, the cone will be preserved in the y-direction, regardless of the magnitude of b. In this gauge, we can write Eq. (4.1.8) as follows

$$\begin{aligned}\Big[-\partial_z^2 + (\kappa_x + bz)^2 + \kappa_y^2 + b\tau_0 \otimes \sigma_y + 1 - \varepsilon^2 + U(z) \\ - \mathrm{i}\, f\alpha_z + 2\varepsilon f z - f^2 z^2\Big]\chi(z) = 0\,,\end{aligned} \tag{4.3.4}$$

where we have already exploited translational symmetry in the XY-plane. Let

$$\mu = (b^2 - f^2)^{1/4}\,. \tag{4.3.5}$$

Hereafter we shall consider $b, f > 0$. Then, we can introduce

$$s = -\sqrt{2}\mu\,(z - z_0)\,, \tag{4.3.6a}$$

$$p = \frac{1}{2\mu^2}\left[\varepsilon^2 - 1 - \kappa^2 + \mu^4 z_0^2\right]\,, \tag{4.3.6b}$$

where

$$z_0 = -\frac{\kappa_x b + \varepsilon f}{\mu^4} . \tag{4.3.7}$$

As a result, Eq. (4.3.4) can be written as follows

$$\left[-\partial_s^2 + \frac{s^2}{4} - p + \mathcal{M}\right] \chi(s) = \delta(s - s_0)\mathcal{N}\chi(s) , \tag{4.3.8}$$

where $s_0 = s(z = 0)$ and

$$\mathcal{M} = \frac{\mathrm{i}}{2\mu^2}(b\alpha_x - f)\,\alpha_z , \qquad \mathcal{N} = \mathrm{i}\,\frac{\sqrt{2}}{\mu}\alpha_z\beta . \tag{4.3.9}$$

Since we shall be interested in treating the term on the right as a perturbation, it is convenient to diagonalize the matrix $\mathcal{M}$, which has eigenvalues $\pm 1/2$, each of which is doubly degenerate. We can then find a matrix U such that $U\mathcal{M}U^{-1} = \beta/2$. If we do so, Eq. (4.3.8) transforms to

$$\left[-\partial_s^2 + \frac{s^2}{4} - p + \frac{1}{2}\beta\right] \mathbf{\Phi}(s) = \delta(s - s_0)\mathcal{W}\mathbf{\Phi}(s) , \tag{4.3.10}$$

where $\mathcal{W} = U\mathcal{N}U^{-1}$ and $\mathbf{\Phi}(s) = U\chi(s)$. We will solve this problem using Green's functions, as in the previous chapter. Treating the term on the right of the previous equation as a perturbation, we can use the Lippman-Schwinger equation to write

$$\mathbf{\Phi}(s) = G(s, s_0)\mathcal{W}\mathbf{\Phi}(s_0) . \tag{4.3.11}$$

The Green's function satisfies

$$\left[-\partial_s^2 + \frac{s^2}{4} - p + \frac{1}{2}\beta\right] G(s, s') = \delta(s - s_0)\mathbb{1}_4 . \tag{4.3.12}$$

Since Eq. (4.3.11) has to hold for all s, it has to hold in particular for $s = s_0$ and, in order to have non-trivial solutions, we must ask

$$\det\left[\mathbb{1}_4 - G(s_0, s_0)\mathcal{W}\right] = 0 . \tag{4.3.13}$$

Taking into account that, formally, $G(s, s')$ represents the inverse of the operator acting on it and in this case the latter is diagonal, we can propose for $G(s, s')$ to be diagonal as well

$$G(s, s') = \begin{pmatrix} g_+(s, s') & 0 \\ 0 & g_-(s, s') \end{pmatrix} \otimes \mathbb{1}_2 . \tag{4.3.14}$$

As a result, it is straightforward to show that Eq. (4.3.13) implies that

$$g_+(s_0, s_0)g_-(s_0, s_0) = \frac{\mu^2}{2} , \tag{4.3.15}$$

and $g_\pm(s, s')$ satisfies

$$\left[-\partial_s^2 + \frac{s^2}{4} - p_\pm\right] g_\pm(s, s') = \delta(s - s_0) , \tag{4.3.16}$$

with $p_\pm = p \mp 1/2$. This is the equation for the Green's function of a (non-relativistic) harmonic oscillator only if $b > f$, in which case s is real. In fact, this regime is the actual regime that can be assessed using the homogeneous Lippman-Schwinger equation since this situation leads to bound states with $G(s, s') \to 0$ as $|s|, |s'| \to \infty$. Indeed, in the absence of electric field, we expect to obtain Landau levels at a distance $z_0 = -\kappa_x \ell_B^2/d^2$ from the interface. Notice that this is exactly the same that one would obtain for a two-dimensional electron gas in the Landau gauge. That is, the two coordinates perpendicular to the magnetic field become linked by relating the momentum of one to the position of the orbit in the other. As we shall see, however, in contrast to the case of Landau levels in a two-dimensional electron gas, the energies do depend on κ_x, becoming dispersive. Although z_0 provides a measure of the distance of the Landau orbits to the interface, ℓ_B provides a measure of the extent of such orbits. Hence, in order to assess the effects of the boundary, one would require $z_0 \to 0$, equivalently $\kappa_x \to 0$, and ℓ_B to be larger than d. A detailed account of other regimes in the absence of electric field is described in reference [24]. Finally, it is interesting to observe that the regime we are considering corresponds to $\ell_F > \ell_B > d$, where ℓ_F is the electric length that was introduced in the previous chapter. That is, we require the electric field potential to be smooth enough so as to be almost constant on each Landau orbit, although affecting them separately.

The Green's function for the harmonic oscillator is known and can be found in Refs. [25, 26]. In our case, we have an oscillator of mass 1/2 and frequency 1, which leads to

$$g_\pm(s, s') = \frac{1}{\sqrt{2\pi}}\Gamma\left(\frac{1}{2} - p_\pm\right) D_{p_\pm - 1/2}(s_>) D_{p_\pm - 1/2}(-s_<) , \tag{4.3.17}$$

where $\Gamma(z)$ is the Gamma function, $D_\gamma(z)$ is the parabolic-cylinder function [27], $s_> = \max\left(s, s'\right)$ and $s_< = \min\left(s, s'\right)$. Using Eq. (4.3.15), we find an equation for the energies

$$D_p(s_0)D_p(-s_0)D_{p-1}(s_0)D_{p-1}(-s_0) + \frac{\pi\mu^2}{p\Gamma^2(-p)} = 0 . \tag{4.3.18}$$

This equation reduces to the one found by Agassi in reference [24] when $f = 0$. As we can observe in this equation, the energies will only depend on two factors, in agreement to our discussion above. Indeed, on the one hand they will depend on the ratio between the distance to the interface, $z_0 d$, and the magnetic length through

$s_0 = \sqrt{2}z_0 d/\ell_B$. On the other, they will depend on the ratio between the surface state decay length d and ℓ_B, $\mu = d/\ell_B$. It is instructive to consider first the effect of b alone, as in [24]. We do so in what follows.

4.3.1 Topological Protection in the Absence of Electric Field

We shall be interested in $z_0 \to 0$ and $\ell_B > d$ for the reasons stated above, so we will be interested in considering $\kappa_x \to 0$ and $b \to 0$. Let us first discuss however the case of $\kappa_x = 0$, in order to observe if the Dirac cone remains unaltered in the Y-direction regardless of the magnetic field. In this case, Eq. (4.3.18) is written as follows

$$\left[D_p(0)D_{p-1}(0)\right]^2 + \frac{\pi\mu^2}{p\Gamma^2(-p)} = 0\,. \tag{4.3.19}$$

Taking into account that

$$D_p(0) = 2^{p/2}\frac{\sqrt{\pi}}{\Gamma\left(\frac{1}{2}-\frac{p}{2}\right)}\,, \tag{4.3.20}$$

using $\Gamma(z+1) = z\Gamma(z)$ and the Legendre duplication formula [27], $\Gamma(2z) = 2^{2z-1}\Gamma(z)\Gamma(z+1/2)/\sqrt{\pi}$, we obtain

$$\frac{1+2p\mu^2}{p^2\Gamma^2(-p)} = 0\,. \tag{4.3.21}$$

There are now two possibilities: either the denominator goes to infinity or the numerator goes to zero. The denominator goes to infinity whenever p is an integer different from zero. Indeed, $\Gamma(-p)$ is not analytic for $p \in \mathbb{Z}_0^+$. However, $\Gamma(-p) = -1/p + \mathcal{O}(p)$ when $p \to 0$, which implies that the product $p\Gamma(-p) \to -1$ when $p \to 0$. Therefore, the denominator goes to infinity whenever $p \in \mathbb{Z}^+$. Also, $\Gamma(-p) \to \infty$ as $p \to -\infty$, so the denominator also goes to infinity when $z \to -\infty$. The numerator, on the other hand, can only go to zero if $p > 0$. Hence, $p = 0$ cannot be a solution to Eq. (4.3.21). As we said, if $p < 0$, both the numerator and the denominator can lead to Eq. (4.3.21) to be satisfied. However, the numerator can be strictly zero for arbitrary values of magnetic field, whereas $p \to -\infty$ requires $b \to 0$. Therefore, if $p < 0$, we have that $1 + 2p\mu^2 = 0$, which taking into account the definition of p [cf. Eq. (4.3.6b)] implies that

$$\varepsilon_s = s\kappa_y\,, \tag{4.3.22}$$

with $s = \pm$.

As expected, the Dirac dispersion is preserved regardless of the magnetic field. However, our argument in favour of this result was saying that the physics in the direction of the field should be unaffected as occurs in ordinary two-dimensional electron gases. Mathematically, this can be observed in the fact that, if $f = 0$, then s_0 becomes independent of ε and the only dependence is found in p, where we observe $\varepsilon^2 - \kappa_y^2$. There is yet a deeper reasoning behind this result, linked to topology. We know that the Dirac point is protected by time-reversal symmetry, but the magnetic field breaks such a symmetry, so we must look in a different direction instead. One extra symmetry of the bulk that was accounted for in our model although without explicitly discussing it is mirror symmetry. Indeed, if we take a look at the lattice of Bi_2Se_3 [cf. Chap. 2], we can observe that the XZ-plane is a mirror plane. That is, if we take $y \to -y$, the system stays the same. This symmetry is equivalent to the product of inversion, I, and C_2 symmetry about the Y axis, C_2^y. In our model, we asked for both symmetries to hold separately, which implies that, indeed, the Hamiltonian is mirror-symmetric. Moreover, since we kept terms to lowest order in $\boldsymbol{\kappa}$, there is no distinction between the X and Y direction in our model. In fact, our model is also symmetric if we perform a C_2 rotation about the X axis, which is not a true symmetry of the system. Of course, a model that includes more terms would account for the fact that the two directions are not equivalent, leading to hexagonal warping effects [28]. This implies that, in our case, it is not too important whether we apply the magnetic field in the X or Y directions. In fact, in reference [24], the field is applied along the X direction instead, and the same results are obtained. Moreover, our model is actually invariant under continuous rotations about the Z axis, which implies that any plane perpendicular to the XY-plane would be an equally valid candidate for being a mirror plane.[1] To be concrete, let us say that the mirror plane is the XZ-plane, as it is in the real system. Then, the mirror symmetry operator can be written as [cf. Chap. 2]

$$\mathcal{M} = \mathcal{D}(I)\mathcal{D}(C_2^y) = \mathrm{i}\,\tau_z \otimes \sigma_y \,, \tag{4.3.23}$$

which takes $(k_x, k_y, k_z) \to (k_x, -k_y, k_z)$. As we can see, $\mathcal{M}^2 = -\tau_0 \otimes \sigma_0$. That is, similar to time-reversal symmetry, mirror symmetry when considering spin introduces a minus sign when squared. In contrast to time-reversal, it is a unitary symmetry. As a result of squaring to -1, the eigenvalues of $\mathcal{M}$ are $\pm\mathrm{i}$, each doubly degenerate. It is not difficult to see that the bulk Hamiltonian commutes with $\mathcal{M}$ in the mirror plane, that is, when $\kappa_y = 0$. This allows to block-diagonalize the Hamiltonian, each block corresponding to one of the two eigenvalues of $\mathcal{M}$. Let $\boldsymbol{u}_\pm^\alpha$ be the four eigenvectors of $\mathcal{M}$ with eigenvalue $\pm\mathrm{i}$ and $\alpha = 1, 2$. Then, a unitary transformation that brings $\mathcal{M}$ into a diagonal form is

$$U = \left[\boldsymbol{u}_+^1, \boldsymbol{u}_+^2, \boldsymbol{u}_-^1, \boldsymbol{u}_-^2\right] \,. \tag{4.3.24}$$

The operator $\mathcal{M}$ becomes $\mathcal{M} = \mathrm{i}\,\tau_z \otimes \sigma_0$ and the Hamiltonian can be written as follows

[1] In fact, this is what we have exploited when saying that the dispersion depends solely on κ in the absence of fields.

$$\mathcal{H} = -s_\Delta \tau_0 \otimes \sigma_z - \kappa_z \tau_0 \otimes \sigma_x + \kappa_x \tau_z \otimes \sigma_y + \kappa_y \tau_y \otimes \sigma_y \,, \tag{4.3.25}$$

where $s_\Delta = \operatorname{sgn} \Delta$, with Δ the energy gap. It is then clear that if $\kappa_y = 0$, then $[\mathcal{H}, \mathcal{M}] = 0$ and both $\mathcal{H}$ and $\mathcal{M}$ can be simultaneously diagonalized. In fact, the unitary transformation has already brought both $\mathcal{M}$ and $\mathcal{H}$ into diagonal form, the latter when $\kappa_y = 0$. The two blocks of $\mathcal{H}$ are then straightforwardly found to be

$$\mathcal{H}_\eta = -s_\Delta \sigma_z - \kappa_z \sigma_x + \eta \kappa_x \sigma_y \,, \tag{4.3.26}$$

where $\eta = \pm 1$ corresponds to the two subspaces of mirror symmetry defined by its eigenvalues, $\pm \mathrm{i}$. These two blocks correspond to two dimensional massive Dirac Hamiltonians. Such Hamiltonians are employed in describing the physics of so-called Chern insulators. These Chern insulators where originally proposed by Haldane [7] in graphene (by the time called two-dimensional graphite). Haldane's idea was to show that one can obtain quantum Hall physics without Landau levels, the only requirement being breaking time-reversal symmetry. The model, which is absent of any symmetries, belongs to the A class in the Altland-Zirnbauer classification and topologically different ground states are classified by an integer. This integer is known as the Chern number [29]. The Hamiltonian we have found is not exactly that of Haldane, but one due to Qi, Wu and Zhang [30] in a square lattice that also describes the physics of Chern insulators. Depending on the sign of the mass term, the Chern number may be either zero (negative mass) or one (positive mass). In order to see why this is so, one can reguilarize the two-dimensional massive Dirac Hamiltonian in a square lattice

$$\mathcal{H}_L = \boldsymbol{d}(\boldsymbol{\kappa}) \cdot \boldsymbol{\sigma} \,, \tag{4.3.27}$$

where

$$d_x(\boldsymbol{\kappa}) = -\sin(\kappa_z) \,, \tag{4.3.28a}$$
$$d_y(\boldsymbol{\kappa}) = \eta \sin(\kappa_x) \,, \tag{4.3.28b}$$
$$d_z(\boldsymbol{\kappa}) = -s_\Delta - 2 + \cos(\kappa_x) + \cos(\kappa_z) \,. \tag{4.3.28c}$$

A non-zero Chern number is achieved if the vector $\boldsymbol{d}(\boldsymbol{\kappa})$ describes a closed surface enclosing the origin upon varying $\boldsymbol{\kappa}$ through the Brillouin zone [31] [cf. this approach to that utilized for studying the topology of the SSH model and graphene in the second chapter]. If $s_\Delta = 1$, then $d_z(\boldsymbol{\kappa})$ is always negative, $\boldsymbol{d}(\boldsymbol{\kappa})$ cannot enclose the origin and the Chern number is zero. However, if $s_\Delta = -1$, then $\boldsymbol{d}(\boldsymbol{\kappa})$ encloses the origin and the Chern number[2] coincides with η. Therefore, the total Chern number

[2]There is an ambiguity on the sign of the Chern number, so we choose a definition such that it coincides with η. However, it would be equally valid to say that the Chern number is $-\eta$. The only important point is to stick to the chosen definition when calculating Chern numbers in a given model. For concreteness, we use here [31]

obtained by adding up the Chern numbers of the two mirror subspaces is clearly zero in either case, since the two Chern numbers of the two mirror subspaces are equal and opposite. However, it is possible to define a *mirror Chern number* as the difference between Chern numbers of each subspace [32]. More precisely, if $n_{\pm}$ is the Chern number of the mirror subspace $\pm$, the mirror Chern number is usually defined as $n_{\mathcal{M}} = (n_+ - n_-)/2$. In that case, the phase with $s_\Delta = 1$ which has zero Chern number in both subspaces renders a mirror Chern number of zero as well; however, the phase with $s_\Delta = -1$ with Chern numbers of ± 1 for each block leads to a mirror Chern number of 1. Therefore, even though the total Chern number is zero, as expected for a time-reversal-symmetric phase, the two phases are topologically distinct. Moreover, the $\mathbb{Z}_2$ invariant may also be zero, as occurs in the case of SnTe, where there is parity inversion as in Bi_2Se_3 but it occurs at an even number of TRIMs in the Brillouin zone. However, since SnTe displays mirror symmetry, the discussion above is still valid and it can be characterized with a mirror Chern number. Therefore, SnTe is not a topological insulator in the sense that its $\mathbb{Z}_2$ index is zero, but it is a *topological crystalline insulator*, since it displays topological behaviour as a consequence of crystalline symmetries. This is why the model we are using is also valid for such materials [33, 34], as was briefly discussed in the second chapter. Interestingly, Bi_2Se_3 also has mirror symmetry. This implies that time-reversal-symmetry-breaking perturbations that preserve mirror symmetry do not destroy the topology of Bi_2Se_3, despite having a trivial $\mathbb{Z}_2$ invariant. Therefore, Bi_2Se_3 is said to be a dual topological insulator [35], in the sense that it can present both a strong topological insulating regime (non-zero $\mathbb{Z}_2$ invariant), as well as a crystalline topological insulating one. Hence, the topological behaviour of Bi_2Se_3 can still survive even in presence of magnetic perturbations. However, such perturbations have to be such that mirror symmetry is preserved. This means that an external magnetic field has to be perpendicular to the mirror planes, since it is a pseudovector. Since we have found a topologically nontrivial behaviour, from the bulk-boundary correspondence we would expect boundary modes to occur at the interface of a system where Δ changes sign. That is, if we consider Eq. (4.3.26) and have $\Delta = \mathrm{sgn}\,(z)$, then $s_\Delta = \mathrm{sgn}\,(z)$ and we want to solve

$$\mathcal{H}_\eta \psi_\eta(z) = \varepsilon_\eta \psi_\eta(z)\,. \tag{4.3.29}$$

We propose the following *ansatz*: $\psi_\eta(z) \propto \exp(-|z|)\phi$ with ϕ a constant vector. Then, we obtain

$$\left[-\mathrm{sgn}\,(z)\,(\sigma_z + \mathrm{i}\,\sigma_x) + \eta\kappa_x\sigma_y\right]\phi = \varepsilon_\eta\phi\,. \tag{4.3.30}$$

The z-dependent term disappears if we ask

$$n = \frac{1}{4\pi}\int_{\mathbb{T}_2} d^2\kappa\, \widehat{\boldsymbol{d}}(\boldsymbol{\kappa})\cdot\left(\partial_x\widehat{\boldsymbol{d}}(\boldsymbol{\kappa})\times\partial_z\widehat{\boldsymbol{d}}(\boldsymbol{\kappa})\right)\,,$$

where $\widehat{\boldsymbol{d}} = \boldsymbol{d}/|\boldsymbol{d}|$ is a unit vector that traces a closed surface when $\boldsymbol{\kappa}$ is carried through the Brillouin zone torus, $\mathbb{T}_2$.

$$(\sigma_z + \mathrm{i}\,\sigma_x)\,\phi = 0\,, \tag{4.3.31}$$

or, equivalently, we require that ϕ is an eigenstate of σ_y with eigenvalue $+1$. Therefore, upon inserting that eigenstate into Eq. (4.3.30) we trivially obtain

$$\varepsilon_\eta = \eta\kappa_x\,. \tag{4.3.32}$$

For each value of η, this dispersion corresponds to a unidirectional-moving edge mode, since the group velocity is nothing but η. These modes are often referred to as *chiral* and are ubiquitous to quantum Hall effect phenomena. Since we have both chiralities, which are realized in $\eta = \pm 1$, and these stem from mirror symmetry, we refer to such a chirality as a *mirror chirality* [32]. The meaning of this result is that the Dirac cones in our model can be understood either from time-reversal symmetry or from mirror symmetry. Since this behaviour occurs due to having nonzero mirror Chern numbers, we can say that mirror chirality is also a topological quantum number. This implies, as before, that perturbations that do not break mirror symmetry cannot lead to gap openings. In our model, where we kept terms to lowest order in κ, it is unimportant which direction the magnetic field points to, as long as it is contained in the XY plane. However, if more terms are added to the Hamiltonian to include hexagonal warping [28], as mentioned above, then only those directions of the magnetic field that strictly preserve mirror symmetry do not lead to gap openings in the spectrum [35].

After this digression, let us explore the case with $p > 0$. In such a case, we know that the denominator in Eq. (4.3.21) goes to infinity when $p \in \mathbb{Z}^+$, where the numerator stays finite. In such a case, taking into account the definition of p [cf. Eq. (4.3.6b)], we obtain a series of relativistic Landau levels

$$\varepsilon_{n,s} = s\sqrt{1 + 2nb + \kappa_y^2}\,, \qquad n \in \mathbb{Z}^+\,. \tag{4.3.33}$$

These are the Landau levels of the bulk, which appear due to the confinement of the harmonic potential in the growth direction created by the magnetic field. Notice that these are relativistic Landau levels for massive Dirac fermions. However, there is no $n = 0$ Landau level, meaning that the lowest energy is not at $\varepsilon = 1$, as it would be in the bulk, but rather it is higher at $\sqrt{1 + 2b}$. This implies that, upon increasing the magnetic field, the gap effectively widens at the interface. This widening can be quite large. For $b \simeq 0.5$, such a widening is of the order of 40% with respect to the original gap. Agassi [24] interprets this result as the energy cost for an electron in the bulk to cross the boundary (remember that $\kappa_x = 0$ corresponds to $z_0 = 0$).

Let us now explore the case of $\kappa_x \to 0$ and $b \to 0$. We will assume that $\kappa_x \to 0$ sufficiently fast so as to have $s_0 \to 0$. That is, we are considering the Landau orbits to be close to the interface, $z_0 \to 0$, with a spatial ℓ_B extension such that they interact with the boundary. The first thing to notice is that Eq. (4.3.18) is even in s_0, which implies that an expansion around $s_0 = 0$ can only contain even powers of s_0. In fact, performing such an expansion leads to lowest order in s_0^2

$$\frac{1+2p\mu^2}{p^2\Gamma^2(-p)} - \left[D_p^2(0) + pD_{p-1}^2(0)\right]^2 s_0^2 = 0\,, \tag{4.3.34}$$

where we have taken into account Eq. (4.3.21). Notice that the two terms in this equation cannot vanish simultaneously. This is because if $f = 0$, then p becomes independent of κ_x, meaning that the equation above is independent of κ_x except for s_0. If we require the two terms to vanish with $s_0 \neq 0$, we would obtain solutions that are independent of κ_x and, therefore, should hold also for $\kappa_x = 0$. However, those solutions are only provided by the first term and not by both. Therefore, it is the full equation that has to be zero and not each term separately. Taking into account (4.3.20) we can write

$$\left[D_p^2(0) + pD_{p-1}^2(0)\right]^2 = \frac{\pi}{2p\Gamma^2(-p)}\left[\eta(p) + \frac{1}{\eta(p)} + 2\right], \tag{4.3.35}$$

where

$$\eta(p) = \frac{p}{2}\frac{\Gamma^2\left(-\frac{p}{2}\right)}{\Gamma^2\left(\frac{1-p}{2}\right)}\,. \tag{4.3.36}$$

Combined with Eq. (4.3.34) we finally obtain

$$\frac{1}{p^2\Gamma^2(-p)}\left\{1 + 2p\mu^2 - s_0^2 p\left[\eta(p) + \frac{1}{\eta(p)} + 2\right]\right\} = 0\,. \tag{4.3.37}$$

Following a similar comment to the one above, for finite s_0 we must require the term in brackets to be zero, that is,

$$1 + 2p\mu^2 = s_0^2 p\left[\eta(p) + \frac{1}{\eta(p)} + 2\right]. \tag{4.3.38}$$

Taking into account the definition of p, this equation can be written as follows

$$\varepsilon^2 - \kappa_y^2 = \left(\frac{\kappa_x}{b}\right)^2\left(\varepsilon^2 - 1 - \kappa_y^2\right)\left[\eta(p) + \frac{1}{\eta(p)} + 2\right]. \tag{4.3.39}$$

We can now analyze the two regimes discussed when $\kappa_x = 0$ for $b \to 0$. Let us start with the Dirac state. In this case, $\varepsilon^2 - \kappa_y^2 \ll 1$ and $p \simeq -1/2b$. In the limiting case of $b \to 0$, $p \to -\infty$ and we can expand the term in brackets in the equation above

$$\eta(p) + \frac{1}{\eta(p)} + 2 = -\frac{1}{4p^2}\left[1 - \frac{5}{16p^2}\right] + \mathcal{O}(p^{-6})\,, \tag{4.3.40}$$

leading to

$$\varepsilon^2 - \kappa_y^2 \simeq -\frac{\kappa_x^2}{\varepsilon^2 - 1 - \kappa_y^2}\left[1 - \frac{5b^2}{4\left(\varepsilon^2 - 1 - \kappa_y^2\right)^2}\right]. \tag{4.3.41}$$

Since $\varepsilon^2 - \kappa_y^2 \ll 1$, we may approximate the denominators to -1 and finally obtain

$$\varepsilon_s = s\sqrt{\left(1 - \frac{5b^2}{4}\right)\kappa_x^2 + \kappa_y^2}\,, \tag{4.3.42}$$

with $s = \pm$. Interestingly, we have obtained an elliptic Dirac cone which, upon turning off the field, becomes the original Dirac cone. This result implies an anisotropic renormalization of the Fermi velocity. Indeed, the cone is widened in the X-direction, while remaining unaltered in the Y-direction. We shall explore the reasoning behind this result shortly. However, let us first look at the bulk Landau levels. In that case, we can expand around positive integers to write

$$\eta(p) + \frac{1}{\eta(p)} + 2 \simeq \frac{c(n)}{(p-n)^2}\,, \qquad n \in \mathbb{Z}^+\,, \tag{4.3.43}$$

where $c(n)$ is a constant that depends on n and which has no closed expression. A few examples are $c(1) = 2/\pi$, $c(2) = 1/\pi$, $c(3) = 3/2\pi \ldots$ Combined with Eq. (4.3.39) and to lowest order in κ_x, this expansion leads to

$$\varepsilon_s = s\sqrt{1 + 2nb + \kappa_y^2 \pm \sqrt{\frac{8nbc(n)}{1+2nb}}\kappa_x}\,. \tag{4.3.44}$$

As we can observe, the Landau levels split apart when $\kappa_x \neq 0$. In order to understand these results, it pays off to solve Eq. (4.3.18) numerically. The results are shown in Fig. 4.3 for $b = 0.5$.

Let us try to understand these results. For that matter, it is important to bear in mind that the magnetic field links the momentum in the X-direction, κ_x, and the position of the Landau orbits, z_0. Indeed, $z_0 = -\kappa_x \ell_B^2$, with ℓ_B in units of d. Hence, if $\kappa_x = 0$ [cf. panels (a) and (b) of Fig. 4.3], then the orbit center coincides with the position of the topological boundary. As it can be observed, the presence of a Dirac-like state inhibits the existence of a zero Landau level. Moving away from the boundary by increasing κ_x [panels (a) and (c)], the doubly degenerate levels split apart and the Dirac state evolves towards the bulk zero Landau level. Finally, far enough from the boundary [panels (a) and (d)], the Landau levels of the bulk system are recovered, with a singly degenerate zero Landau level, all other levels being doubly degenerate. It is interesting to observe that, similarly to non-relativistic quantum mechanics, Landau levels are only dispersive in the direction parallel to the field when the bulk system is considered. However, the presence of the interface leads to a dispersion and, therefore, to the possibility of defining nonzero group velocities in the in-plane direction perpendicular to the field. As shown in the inset of panel (a), the dispersion

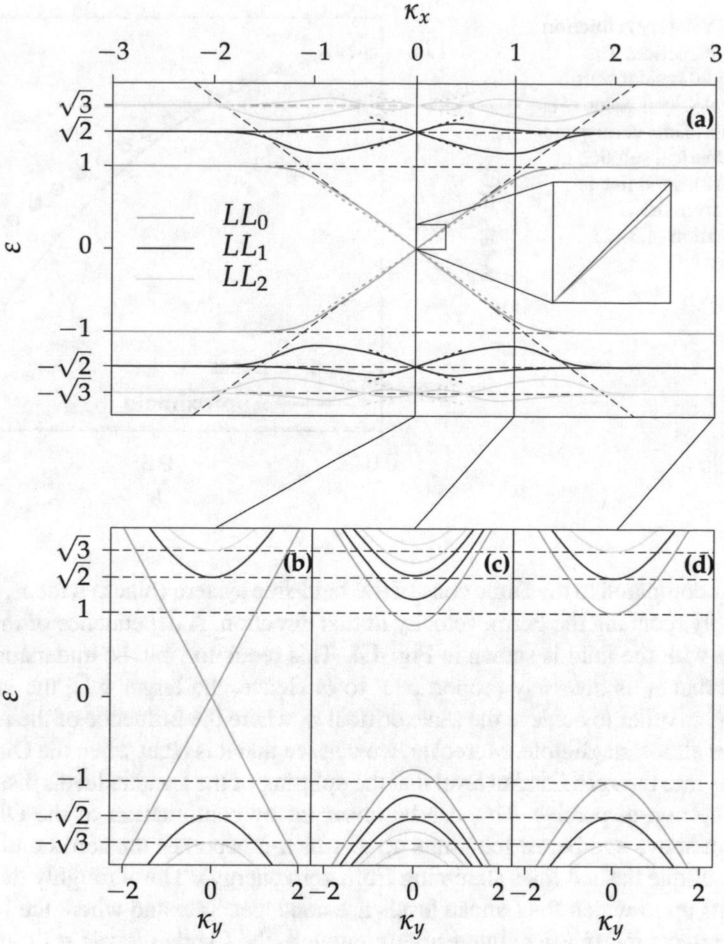

Fig. 4.3 Landau levels and Dirac state in a topological boundary. **a** Energy levels for $\kappa_y = 0$. Close to the boundary, $\kappa_x \to 0$, the Dirac state coexists with dispersive Landau levels. Far from the boundary, the levels evolve into bulk Landau levels, indexed by LL_n. The continued Dirac state (black dashed oblique lines) limit the region where the Landau levels are dispersive. Coloured dashed lines correspond to the approximations (4.3.42) and (4.3.44). The inset shows the widening of the Dirac cone (blue) with respect to the field-free system (black). **b–d** Dispersion for three values of κ_x linked to panel (a). The Dirac state evolves into the zero Landau level, the other Landau levels splitting apart and then rejoining again to recover the double degeneracy in the bulk

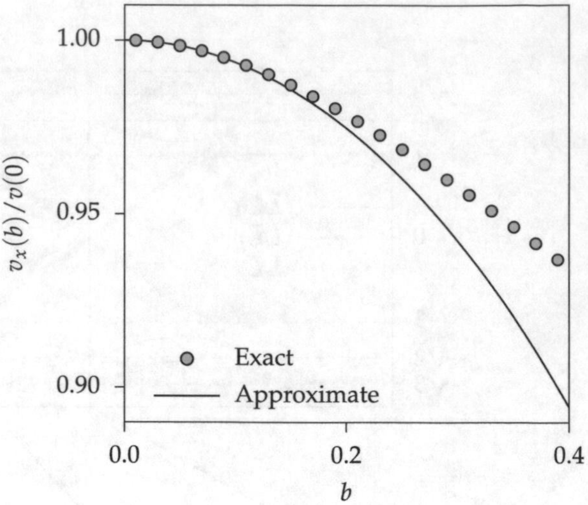

Fig. 4.4 Velocity reduction in the X-direction. The electric field is set to zero and the velocity is reduced as b increases. Dots correspond to the numerical solution of Eq. (4.3.18), solid line is obtained from the approximation (4.3.42)

(blue) as compared to the Dirac cone of the field-free system (black) widens, thereby effectively reducing the Fermi velocity in that direction. A dependence of the Fermi velocity with the field is shown in Fig. 4.4. This reduction can be understood from the fact that z_0 is inversely proportional to b. Hence, the larger b is, the larger κ_x has to be in order to achieve the same critical z_0 where the influence of the interface becomes almost negligible. Moreover, we can see that it is right when the Dirac state saturates into the zero Landau level that the splitting of the Landau levels disappears, becoming non-dispersive. This can be observed by continuation of the Dirac-like spectrum above and below the limits up to which it becomes the zero Landau level (black oblique dashed lines stemming from zero energy). These roughly determine the limits up to which the Landau levels are nondegenerate and where the interface mostly affects the physics. Interestingly enough, the Landau levels split apart and rejoin $2n - 1$ times, with n the Landau index. Possibly this corresponds to avoided crossings between different Landau levels, which in turn obliges these to approach the bulk Landau levels.

4.3.2 Robustness Under Crossed Electric and Magnetic Fields

In this section, we shall include the effect of an electric field. In contrast to the previous section, we have been unable to obtain approximate results. In any case, we would expect from the previous chapter that the Fermi velocity should be reduced isotropically by the electric field. However, one has to be really careful with this picture. Indeed, now κ_x is linked to real space and, therefore, the cone will not widen isotropically since the electric potential is space dependent. We in fact expect that the

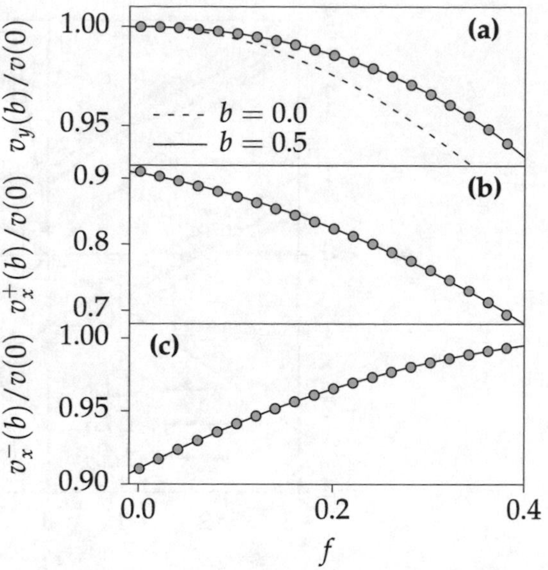

Fig. 4.5 Anisotropic velocity reduction. a The velocity in the Y-direction is reduced by the electric field in the Z-direction in a slower fashion as if there was no magnetic field. **b** The velocity in the positive X-direction is reduced by both the magnetic and the electric fields. **c** The velocity in the negative X-direction increases due to the electric field. In all these figures $b = 0.5$

velocity will depend not only on even powers of f, but also on odd powers. Indeed, since κ_x and position along the growth direction are linked, the asymmetry induced by the electric potential along the Z-direction will be experienced by the cone itself. In Figs. 4.5(b) and (c), we observe this behaviour, where solid lines are nonlinear fits to the data on all powers of f up to f^2. In panel (b), we show the velocity along the positive direction of κ_x, and the opposite in panel (c) for the upper cone. As it can be observed, the velocity becomes highly anisotropic, increasing in the latter situation and decreasing in the former. This occurs due to the tilting of the band edges, as we shall see in a moment. On the other hand, we do expect an isotropic widening along the Y-direction, which is unaffected by the magnetic field. Indeed that happens, as shown in Fig. 4.5(a). The reasoning is the same as in the previous chapter: the Dirac cone widens in order to avoid the surface states from hybridizing with states in the bulk. It is interesting to observe, however, that the reduction is slower than that predicted without the magnetic field (dashed line), which has to do with the fact that the bulk levels are affected by the magnetic field, as we know.

The explanation for the changes in velocity along the X-direction is more clearly understood by observing the energy levels as a function of κ_x in Fig. 4.6 for $b = 0.5$ and $f = 0.3$. By tilting the band edges, both the Landau levels and the Dirac state tilt in response to such tiltings. In fact, this tilting is such that the bulk Landau levels far from the boundary disperse exactly with the usual drift velocity $\boldsymbol{E} \times \boldsymbol{B}/B^2$, which in this case would be $-(f/b)\,\hat{\boldsymbol{x}}$. This can be thought of as if each Landau level experienced a local potential, that allows to still define the Landau levels locally, while only shifting them in energy. This is what we would expect, as discussed earlier in the text, in the regime where $\ell_F > \ell_B > d$, such that the electric field is smooth enough to be almost constant for each Landau level. Interestingly enough, the

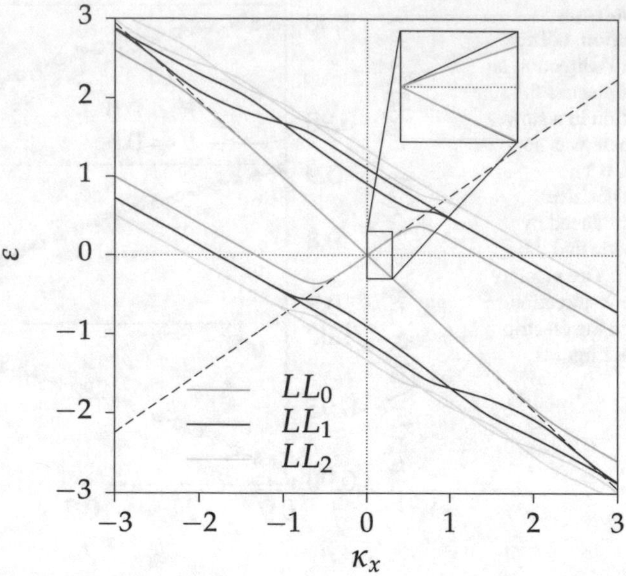

Fig. 4.6 Landau levels tilted by the electric field. In the figure, the first few Landau levels are shown, tilted by the electric field. Far from the boundary (large κ_x), the bulk Landau levels disperse with the usual velocity $-(f/b)\widehat{x}$. The splitted parts of the Landau levels are contained within the anisotropic cone defined by the dashed lines, which are a continuation of the Dirac-like states. The inset shows the anisotropic dispersion (blue) as compared to the field-free system (black). In this figure, $b = 0.5$ and $f = 0.3$

number of crossings within each Landau level is still given by $2n - 1$ and these are still contained in what would be the continued Dirac regime, as shown by dashed lines. Also, different Landau levels do not hybridize. This can be explained by thinking about slowly moving away from the boundary: if hybridizations were allowed, it would not be possible to trace the evolution of Landau levels to the doubly degenerate Landau levels in the bulk, just like in the system without electric field. The anisotropy in the velocity along the X-direction also becomes clear now. Indeed, since the zero Landau levels (positive and negative) also have group velocity $-(f/b)\widehat{x}$ when far from the boundary, then both levels must become parallel at some point in space. As a result, the Dirac cone will have to curve and it will do so anisotropically. It is also interesting to point out that the Dirac point remains robust also in this case, since an electric field pointing in the Z-direction does not break mirror symmetry.

4.4 Conclusions

In this chapter, we have discussed the effect of magnetic and electric fields in a topological boundary. On the one hand, we have been able to observe that, in the case of fields perpendicular to the boundary, the surface Dirac cone evolves into the usual relativistic Landau levels with a renormalized velocity as that given in the previous chapter. On the other hand, we have discussed that, even though time-reversal symmetry is broken by the presence of a magnetic field, our system is mirror symmetric and belongs to the *topological crystalline insulator* class. This class is absent in the Altland-Zirnbauer classification since such classification only considers the fundamental time-reversal, particle-hole and chiral symmetries. However, this classification can be enriched by exploiting other symmetries, as discussed. We have observed that, in the absence of electric field, the velocity renormalizes in the perpendicular direction of the field. Also, we have observed that, due to the interaction with the boundary, Landau levels behave very differently from those in the bulk, splitting and dispersing. In the bulk, these levels are doubly degenerate and nondispersive, except for the zero Landau level which is singly degenerate, as we know. However, close to the interface, this degeneracy breaks and the levels become dispersive, allowing to define the aforementioned velocity. Finally, upon applying the electric field, the velocity renormalizes as well, although in a very anisotropic fashion. Indeed, in the positive X-direction it lowers, while it increases in the negative X-direction, so that the velocity of the bulk levels approach the usual drift velocity, $-(f/b)\,\widehat{\boldsymbol{x}}$. The velocity in the Y-direction decreases the same in both directions, in accordance to the results of the previous chapter. However, this reduction is slower than in the absence of electric field due to the effect of the magnetic field on the bulk levels. In short, we can conclude that the physics of a topological boundary in electric and magnetic fields is not at all trivial. In fact, we have observed that there are a number of different effects that can occur and differ from the bulk system, as we have recapitulated above.

References

1. Kv Klitzing, Dorda G, Pepper M (1980) New Method for High Accuracy Determination of the Fine-Structure Constant Based on quantized hall resistance. Phys Rev Lett 45(494):209
2. Tsui DC, Stormer HL, Gossard AC (1982) Two-dimensional magneto transport in the extreme quantum limit. Phys Rev Lett 48:1559
3. Laughlin RB (1981) Quantized Hall conductivity in two dimensions. Phys Rev B 23:5632
4. Thouless DJ, Kohmoto M, Nightingale MP, den Nijs M (1982) Quantized Hall conductance in a two-dimensional periodic potential. Phys Rev Lett 49:405
5. Niu Q, Thouless DJ, Wu Y-S (1985) Quantized Hall conductance as a topological invariant. Phys Rev B 31:3372
6. Niu Q, Thouless DJ (1987) Quantum Hall effect with realistic boundary conditions. Phys Rev B 35:2188
7. Haldane FDM (1988) Model for a quantum Hall effect without Landau levels: condensed-matter realization of the parity anomaly. Phys Rev Lett 61:2015
8. Altland A, Simons BD (2010) Condensed matter field theory. Cambridge University Press, Cambridge
9. Dean CR, Wang L, Maher P, Forsythe C, Ghahari F, Gao Y, Katoch J, Ishigami M, Moon P, Koshino M, Taniguchi T, Watanabe K, Shepard KL, Hone J, Kim P (2013) Hofstadter's butterfly and the fractal quantum Hall effect in Moiré superlattices. Nature 497:598
10. Ponomarenko LA, Gorbachev RV, Yu GL, Elias DC, Jalil R, Patel AA, Mishchenko A, Mayorov AS, Woods CR, Wallbank JR, Mucha-Kruczynski M, Piot BA, Potemski M, Grigorieva IV,

Novoselov KS, Guinea F, Fal'ko VI, Geim AK (2013) Cloning of Dirac fermions in graphene superlattices. Nature 497:594

11. Hunt B, Sanchez-Yamagishi JD, Young AF, Yankowitz M, LeRoy BJ, Watanabe K, Taniguchi T, Moon P, Koshino M, Jarillo- Herrero P (2013) Ashoori RC (2013) Massive dirac fermions and hofstadter butterfly in a van der Waals heterostructure. Science 340:1427
12. Jaksch D, Zoller P (2003) Creation of effective magnetic fields in optical lattices: the Hofstadter butterfly for cold neutral atoms. New J Phys 5:56
13. Lin Y-J, Compton RL, Jiménez-Garcia K, Porto JV, Spielman IB (2009) Synthetic magnetic fields for ultracold neutral atoms. Nature 462:628
14. Aidelsburger M, Atala M, Lohse M, Barreiro JT, Paredes B, Bloch I (2013) Realization of the hofstadter hamiltonian with ultracold atoms in optical lattices. Phys Rev Lett 111:185301
15. Miyake H, Siviloglou GA, Kennedy CJ, Burton WC (2013) Ketterle W (2013) Realizing the harper hamiltonian with laser-assisted tunneling in optical lattices. Phys Rev Lett 111:185302
16. Hofstadter DR (1976) Energy levels and wave functions of Bloch electrons in rational and irrational magnetic fields. Phys Rev B 14:2239
17. Castro Neto AH, Guinea F, Peres NMR, Novoselov KS (2009) Geim AK (2009) The electronic properties of graphene. Rev Mod Phys 81:109
18. Goerbig MO (2011) Electronic properties of graphene in a strong magnetic field. Rev. Mod. Phys. 83:1193
19. Cheng P, Song C, Zhang T, Zhang Y, Wang Y, Jia J-F, Wang J, Wang Y, Zhu B-F, Chen X, Ma X, He K, Wang L, Dai X, Fang Z, Xie X, Qi X-L, Liu C-X, Zhang S-C, Xue Q-K (2010) Landau quantization of topological surface states in Bi_2Se_3. Phys Rev Lett 105:076801
20. Hanaguri T, Igarashi K, Kawamura M, Takagi H, Sasagawa T (2010) Momentum-resolved Landau-level spectroscopy of Dirac surface state in Bi_2Se_3. Phys Rev B 82:081305
21. Rammal R (1985) Landau level spectrum of Bloch electrons in a honeycomb lattice. J Phys France 46:1345
22. Lukose V, Shankar R, Baskaran G (2007) Novel electric field effects on landau levels in graphene. Phys Rev Lett 98:116802
23. Peres NMR, Castro EV (2007) Algebraic solution of a graphene layer in transverse electric and perpendicular magnetic fields. J Phys: Condens Matter 19:406231
24. Agassi D (1994) Landau levels in a band-inverted junction and quantum well. Phys Rev B 49:10393
25. Domínguez-Adame F (1991) Spectroscopy of a perturbed dirac oscillator. Europhys Lett 15:569
26. Glasser ML, Nieto LM (2015) The energy level structure of a variety of one-dimensional confining potentials and the effects of a local singular perturbation. Can J Phys 93:1588
27. Abramowitz M, Stegun I (1972) Handbook of mathematical functions. Dover, New York
28. Fu L (2009) Hexagonal warping effects in the surface states of the topological insulator Bi_2Te_3. Phys Rev Lett 103:266801
29. Nakahara M (2003) Geometry, topology and physics. Taylor & Francis, Boca Raton, USA
30. Qi X-L, Wu Y-S, Zhang S-C (2006) Topological quantization of the spin Hall effect in two-dimensional paramagnetic semiconductors. Phys Rev B 74:085308
31. Hasan MZ, Kane CL (2010) Colloquium: topological insulators. Rev Mod Phys 82:3045
32. Teo JCY, Fu L, Kane CL (2008) Surface states and topological invariants in three-dimensional topological insulators: application to $Bi_{1-x}Sb_x$. Phys Rev B 78: 045426 (2008)
33. Hsieh TH, Lin H, Liu J, Duan W, Bansil A, Fu L (2012) Topological crystalline insulators in the SnTe material class. Nat Commun 3:982
34. Ando Y, Fu L (2015) Topological crystalline insulators and topological superconductors: from concepts to materials. Ann Rev Condens Matter Phys 6:361
35. Rauch T, Flieger M, Henk J, Mertig I, Ernst A (2014) Dual topological character of chalcogenides: theory for Bi_2Te_3. Phys Rev Lett 112:016802

Chapter 5
Surface States in δ-doped Topological Boundaries

We have discussed the stability of topological surface states in Bi_2Se_3 and other materials alike against external perturbations. However, when exposed to environmental conditions and room temperature, the fate of these states may not be so clear. Indeed, the presence of impurities would ruin the properties of a trivial semiconductor leading to backscattering processes. However, topological surface states are robust against backscattering and this should not be a problem. Nevertheless, such impurities would free electrons (if such impurities are donors), thereby leading to a leftover of positively charged ions at the surface which will, in turn, create a built-in electric field. Such an electric field will bend the band edges and, together with surface confinement, it will form a quantum well for states in the bulk. As a consequence, a two-dimensional electron gas will form at the surface. The natural question to ask is whether the topological surface state can coexist with the quantum confined bulk states and, if so, how are some essential properties modified with respect to not having a topological surface state.

There has been notable experimental progress in this direction and theoretical modeling has been put forward to explain such experiments. On the one hand, experiments [1–7] based on angle resolved photoemission spectroscopy (ARPES) discuss that upon cleaving Bi_2Se_3 and Bi_2Te_3 in different environments (vacuum chambers, water vapor, air, N_2, O_2 and CO atmospheres) or upon intentionally doping with donors such as Fe and Cu, the Dirac state can actually coexist with the two-dimensional electron gas. Moreover, the latter shows a clear Rashba spin-split dispersion as a consequence of the potential gradient and inversion symmetry breaking at the vacuum-material interface. Such dispersion is predicted to have a large effect on spintronic devices due to the large spin-orbit interactions of the materials at play. The surface state cannot be split, however, since it is nondegenerate, as we already know. On the other hand, theoretical calculations [1, 3–6] rely mostly on *ab initio* calculations that consider coupled Schrödinger-Poisson schemes and density functional techniques, together with tight-binding modeling.

This approach of relying on impurities stuck on the cleaved surface is far from controllable. As an alternative, we propose to evaporate a thin layer of donor atoms

Á. Díaz Fernández, *Reshaping of Dirac Cones in Topological Insulators and Graphene*, Springer Theses, https://doi.org/10.1007/978-3-030-61555-0_5

during growth, forming a δ-layer of impurities at the topological boundary (see Refs. [8, 9]). Instead of applying numeric machinery, we shall consider the Thomas-Fermi approximation [10, 11], which will provide with an exactly solvable model where the essential features of the experiments are obtained.

5.1 Thomas-Fermi Approximation

First of all, it must be remarked that this section is not at all a review on the Thomas-Fermi approximation, for which the reader is referred to Refs. [12, 13]. Instead, it focuses on the exactly solvable problem of a δ-layer, discussed by Ioratti [14]. Nevertheless, it is interesting to briefly state the main ideas. Donor impurities free electrons, leaving behind a positively charged background. However, being negatively charged, such electrons will roam around this positive background, screening the Coulomb potential of the donors. As a result, in the single-particle description, an electron will experience a mean field caused by the donors and its neighbouring electrons. Let $\rho_i(\boldsymbol{r})$ and $\rho_e(\boldsymbol{r})$ be the ionic and electronic charge distributions. As we said, electrons will gather close to the donors, which means that $\rho_i(\boldsymbol{r})$ creates a one-electron confinement potential energy $U(\boldsymbol{r})$. On the other hand, there are electron-electron interactions which, in the single-particle approximation, amounts to a potential $V_H(\boldsymbol{r})$ created by $\rho_e(\boldsymbol{r})$ experienced by a single electron. Such a potential is the Hartree potential. However, such an electron-electron interaction only accounts for electrostatics. One should include in the theory the fact that electrons are fermions and should obey Fermi-Dirac statistics. In doing so, one can work in the Hartree-Fock approximation or, rather, in the density functional scheme with the inclusion of exchange-correlation terms. The Thomas-Fermi approximation neglects such a contribution and we will therefore not include it as well.[1] If we denote the total potential in the single-particle description by $V(\boldsymbol{r}) = U(\boldsymbol{r}) + V_H(\boldsymbol{r})$, then $(1/e)\nabla V(\boldsymbol{r})$ would be the aforementioned mean field experienced by a single electron. Denoting the total charge density as $\rho(\boldsymbol{r}) = \rho_i(\boldsymbol{r}) + \rho_e(\boldsymbol{r})$, then $V(\boldsymbol{r})$ is a solution to the Poisson equation

$$\nabla^2 V(\boldsymbol{r}) = \frac{e\rho(\boldsymbol{r})}{\epsilon} , \tag{5.1.1}$$

where ϵ is the dielectric permittivity of the host material where impurities are introduced into. One requirement that $\rho(\boldsymbol{r})$ has to fulfill and which, in turn, imposes a boundary condition on the Poisson equation, is charge neutrality. That is,

$$\int \mathrm{d}^3\boldsymbol{r}\ \rho(\boldsymbol{r}) = 0 . \tag{5.1.2}$$

[1]Notice that this makes sense in the high-density limit, where the kinetic energy dominates over the interactions [13]. This will be the regime of our interest, as we shall see.

In order to make further progress, we will consider the positively charged background created by the ions to be what is commonly referred to as a *jellium* [13]. That is, it approximates $\rho_i(\boldsymbol{r})$ to an average value. In the case of a δ layer, all impurities are concentrated within the $z = 0$ plane. Denoting by n_S the areal density of impurities, then we can write in the jellium model

$$\rho_i(z) = en_S\delta(z) . \tag{5.1.3}$$

Notice that we are assuming that impurities are donors, that all impurities are ionized and that each impurity contributes with one electron. Also, it is important to observe that this approximation only works in the limit of a large concentration of dopants, so that their discrete distribution is unimportant. On the other hand, the mean field potential will create distortions in the electron density of the otherwise homogeneous electron gas, leading to an inhomogeneous electron gas. However, Thomas and Fermi argue that, if such distortions occur in length scales much larger than the Fermi wavelength, then one cannot locally distinguish the inhomogeneous electron gas from the homogeneous one [10, 11, 13]. Therefore, one would expect the system in equilibrium to have a constant Fermi energy, E_F. In order to achieve such a situation, let us imagine a distorted landscape that we want to fill with water. In order for water to be flat in such a landscape, more water will be found at the valley regions with respect to the amount encountered in the mountain regions. In our case, such a landscape is an energy landscape that we have to fill with electrons and the flatness condition refers to the requirement of having a constant Fermi energy. Regions where $V(\boldsymbol{r})$ is negative are valleys, and otherwise for mountains. Hence, the electron density $n(\boldsymbol{r})$ at a given position $\boldsymbol{r}$ would be obtained by filling up all states of the local Fermi gas from $V(\boldsymbol{r})$ up to E_F. That is, if we denote by $\mathcal{D}(E)$ the three-dimensional density of states, then [15]

$$n(\boldsymbol{r}) = \int_{E_c+V(\boldsymbol{r})}^{E_F} \mathrm{d}E\; \mathcal{D}(E) , \tag{5.1.4}$$

where E_c is the conduction band edge, which is usually set to zero but we will leave it for convenience, as we shall see. In other words, the Thomas-Fermi approximation considers a free electron gas with a spatially dependent band edge. The power of this method is that Eq. (5.1.4) provides an equation for $n(\boldsymbol{r})$ in terms of $V(\boldsymbol{r})$. This allows to effectively decouple the Schrödinger (Dirac) and Poisson equations. Indeed, normally one would have to follow a self-consistent approach:

(1) propose a trial $V(\boldsymbol{r})$,
(2) solve the Schrödinger equation with such $V(\boldsymbol{r})$,
(3) obtain $n(\boldsymbol{r})$ from the many-body wavefunction,
(4) solve the Poisson equation for such $n(\boldsymbol{r})$ to obtain another $V'(\boldsymbol{r})$,
(5) if $V'(\boldsymbol{r})$ converges to $V(\boldsymbol{r})$, then the algorithm is finished and the electron density is $n(\boldsymbol{r})$; else, run through steps (2) – (5) with $V(\boldsymbol{r}) \to V'(\boldsymbol{r})$.

In the Thomas-Fermi approximation, we do not go through such process, but rather provide an approximation for $n(\boldsymbol{r})$ to obtain the potential $V(\boldsymbol{r})$ from the Poisson equation and then plug it into the one-electron Hamiltonian. In any case, taking into account that the Fermi energy of a free electron gas is given by

$$E_F = \frac{1}{2m^*}(3\pi^2 n)^{2/3}\,, \tag{5.1.5}$$

we can see that Eq. (5.1.4) will modify this result to give

$$E_F = \frac{1}{2m^*}\left[3\pi^2 n(\boldsymbol{r})\right]^{2/3} + E_c + V(\boldsymbol{r})\,. \tag{5.1.6}$$

Here m^* is the effective mass. Hence, we can obtain an equation for $n(\boldsymbol{r})$ in terms of $V(\boldsymbol{r})$

$$n(\boldsymbol{r}) = \frac{1}{3\pi^2}\left\{2m^*\left[E_F - E_c - V(\boldsymbol{r})\right]\right\}^{3/2}. \tag{5.1.7}$$

Since $\rho_e(\boldsymbol{r}) = -en(\boldsymbol{r})$, we can write the Poisson equation as follows,

$$\nabla^2 V(\boldsymbol{r}) = \frac{e^2}{\epsilon}\left\{n_S\delta(z) - \frac{1}{3\pi^2}\left\{2m^*\left[E_F - E_c - V(\boldsymbol{r})\right]\right\}^{3/2}\right\}. \tag{5.1.8}$$

In our problem, since we have considered the ionic distribution to be uniform in the XY-plane, the electron density $n(\boldsymbol{r})$ can only depend on z, $n(z)$, which implies that $V(\boldsymbol{r})$ is only dependent on z as well and we can write (5.1.8) as follows

$$\frac{\partial^2 V(z)}{\partial z^2} = \frac{e^2}{\epsilon}\left\{n_S\delta(z) - \frac{1}{3\pi^2}\left\{2m^*\left[E_F - E_c - V(z)\right]\right\}^{3/2}\right\}. \tag{5.1.9}$$

As we can see, we have indeed succeeded in obtaining an equation for $V(z)$ that is decoupled from $n(\boldsymbol{r})$ and, therefore, from the Schrödinger (Dirac) equation. This equation can be solved to give [14]

$$V(z) = E_F - E_c - \frac{\gamma^2}{(\gamma|z|/a^* + \omega)^4}\mathrm{Ry}^*\,, \tag{5.1.10}$$

where

$$\gamma = \frac{2}{15\pi}\,, \quad \omega = \left(\frac{\gamma^3}{\pi n_S (a^*)^2}\right)^{1/5}\,, \tag{5.1.11}$$

being Ry^* and a^* the effective Rydberg and Bohr radius, respectively, which are given by

$$\mathrm{Ry}^* = \frac{1}{2m^*(a^*)^2}\,, \qquad a^* = \frac{4\pi\epsilon}{e^2 m^*}\,. \tag{5.1.12}$$

Notice that this potential is regularized at the origin thanks to the ω term in the denominator and decays much faster than the bare Coulomb potential, as expected from the previously discussed screening. As examined in detail by Ioratti [14], this potential describes neutral structures whenever $E_F = E_c$, so that $V(z)$ vanishes as $|z| \to \infty$. A one-band approximation (Ben Daniel-Duke [16]) with this potential admits exact solutions in terms of Mathieu's functions [17]. However, as pointed out also in Ref. [14], finding the energy spectrum is extremely complex in this scenario. Therefore, a two-band approximation (Dirac) with this potential is out of question. It is then compelling to consider an approximate form for the potential that satisfies the same boundary conditions as (5.1.10). Since we shall be interested in introducing the resulting potential into the Dirac equation, it is interesting to express distances in units of $d = v_z/\Delta$ and energies in units of Δ. Hence, we write

$$\frac{\partial^2 v(\xi)}{\partial \xi^2} = \frac{e^2}{\epsilon} \left\{ \frac{n_S d}{\Delta} \delta(\xi) - \frac{\Delta^{1/2}}{3\pi^2} \left\{ 2m^* \left[-v(\xi) \right] \right\}^{3/2} \right\} . \tag{5.1.13}$$

In this equation, $v = V/\Delta$ and $\xi = z/d$. Thus,

$$\frac{\partial^2 v(\xi)}{\partial \xi^2} = 8\pi \mathrm{Ry}_\Delta N_S \delta(\xi) - \frac{8}{3\pi} \frac{1}{a_\Delta^2 \mathrm{Ry}_\Delta^{1/2}} \left[-v(\xi) \right]^{3/2} . \tag{5.1.14}$$

where $\mathrm{Ry}_\Delta = \mathrm{Ry}^*/\Delta$ is the effective Rydberg in units of Δ, $a_\Delta = a^*/d$ is the effective Bohr radius in units of d and $N_S = a_\Delta n_S d^2$ is the number of impurities in a square plaquette of side $\sqrt{a_\Delta} d$. Integrating this equation around $\xi = 0$, we get the following boundary condition

$$\left. \frac{\partial v(\xi)}{\partial \xi} \right|_{0+} - \left. \frac{\partial v(\xi)}{\partial \xi} \right|_{0-} = 8\pi \mathrm{Ry}_\Delta N_S , \tag{5.1.15}$$

which implies that the electric field at the boundary is discontinuous. However, it is a finite discontinuity, in contrast to the electric field due to the Coulomb potential, leading to a regularization at the origin. The other boundary condition is obtained integrating in all the real line. In order to do so, let us rewrite the charge neutrality condition. Since $\rho(z) = e n_S \delta(z) - e n(z)$, we can write

$$\int_{-\infty}^{\infty} \mathrm{d}z \, n(z) = n_S . \tag{5.1.16}$$

If we take into account that

$$(a^*)^3 n(z) = \frac{1}{3\pi^2} \frac{1}{\mathrm{Ry}_\Delta^{3/2}} \left[-v(z) \right]^{3/2} , \tag{5.1.17}$$

then, the charge neutrality condition is written as

$$\int_{-\infty}^{\infty} \mathrm{d}\xi \, \left[-v(z) \right]^{3/2} = 3\pi^2 N_S a_\Delta^2 \mathrm{Ry}_\Delta^{3/2} . \tag{5.1.18}$$

It is then clear that by integrating equation (5.1.14) in the real line we obtain

$$\left.\frac{\partial v(\xi)}{\partial \xi}\right|_{\infty} - \left.\frac{\partial v(\xi)}{\partial \xi}\right|_{-\infty} = 0\,, \tag{5.1.19}$$

which implies that the electric field has to vanish at $\pm\infty$. Indeed, the symmetry of the problem implies that $v(\xi)$ is even, which means that the two terms above have to vanish identically. A clever choice of a potential that satisfies the boundary conditions is a one-dimensional Yukawa potential [18, 19],

$$v_{\mathrm{app}}(\xi) = -v_0 \exp\left(-\frac{|\xi|}{\eta}\right), \tag{5.1.20}$$

where the two constants $v_0, a > 0$ have to be chosen to satisfy the boundary conditions and we are assuming neutral structures (see above). Notice that this indeed corresponds to a Yukawa potential since, upon Fourier transforming, the Fourier components are of the form $\tilde{v}(q) \propto (q^2 + \eta^{-2})^{-1}$. The discontinuity at the origin requires that

$$v_0 = 4\pi\eta \mathrm{Ry}_{\Delta} N_S\,. \tag{5.1.21}$$

On the other hand, the potential already decays at $\pm\infty$ (assuming $a > 0$). Therefore, in order to obtain another equation that relates v_0 and η, we introduce $v(\xi)$ into the charge neutrality condition (5.1.18), which leads to

$$v_0^{3/2}\frac{4\eta}{3} = 3\pi^2 N_S a_{\Delta}^2 \mathrm{Ry}_{\Delta}^{3/2}\,. \tag{5.1.22}$$

Combining these two equations, we finally obtain

$$\eta = \left(\frac{3^4\pi a_{\Delta}^4}{4^5 N_S}\right)^{1/5} \simeq \frac{3}{4}\left(\frac{a_{\Delta}^4}{N_S}\right)^{1/5}, \tag{5.1.23a}$$

$$v_0 = \left(3^4\pi^6 a_{\Delta}^4 N_S^4\right)^{1/5} \mathrm{Ry}_{\Delta} \simeq 3\pi\,(a_{\Delta}N_S)^{4/5}\,\mathrm{Ry}_{\Delta}\,. \tag{5.1.23b}$$

In order to compare the approximate potential (5.1.20) with the exact potential (5.1.10), we rewrite the exact potential as follows

$$v_{\mathrm{exac}}(\xi) = -v_0\,\frac{c_1}{(c_2 + c_3|\xi|/\eta)^4}\,, \tag{5.1.24}$$

where $c_1 \simeq 43652.64$, $c_2 \simeq 14.72$ and $c_3 \simeq 3.96$. In Fig. 5.1 we show a comparison of $v_{\mathrm{exac}}(\xi)$ and $v_{\mathrm{app}}(\xi)$ and the agreement is noteworthy.

In the following section, we will solve the topological boundary model with the potential $v_{\mathrm{app}}(\xi)$ in order to explore the physics of the experiments mentioned in the introduction. However, before doing so, it is interesting to have some estimates of v_0 and η in the materials of our interest, namely, three-dimensional topological

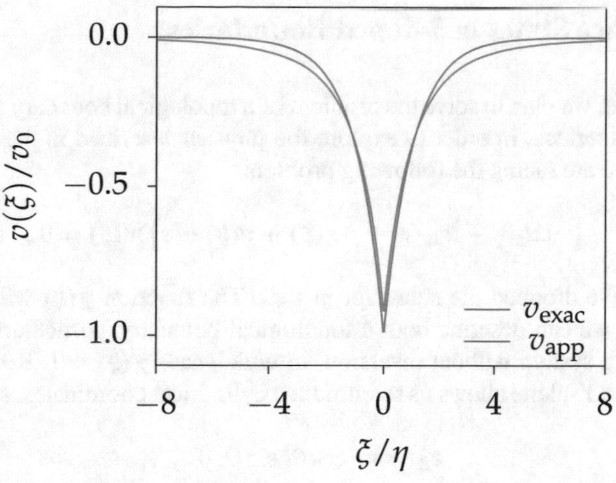

Fig. 5.1 Screened Coulomb potential. A comparison between the exact screened potential, (5.1.24), and the approximate Yukawa potential, (5.1.20), shows very good agreement

insulators and crystalline topological insulators. As a representative of the former, we will take Bi_2Se_3, for which [1, 20] $\Delta \simeq 175$ meV, $v_F \simeq 250$ meV nm, $\epsilon_r \simeq 113$, and $m^* \simeq 0.2m_0$, where ϵ_r is the relative permittivity and m_0 is the bare electron mass.[2] As a result, $d \simeq 1.4$ nm, $a^* \simeq 30$ nm and $\text{Ry}^* \simeq 0.2$ meV. On the other hand, typical parameters of IV-VI semiconductors, such as the topological crystalline insulator SnTe, are [21, 22] $\Delta \simeq 75$ meV, $v_F \simeq 338$ meV nm, $\epsilon_r \simeq 45$ and $m^* \simeq 0.05m_0$. Hence, $d \simeq 4.5$ nm, $a^* \simeq 48$ nm and $\text{Ry}^* \simeq 0.3$ meV. Typically, impurity concentrations are in the range of 10^{10} cm^{-2} to 10^{12} cm^{-2} [19]. Hence, it makes sense to write $n_S = x\ 10^{10}\ \text{cm}^{-2} = 10^{-4}\ \text{nm}^{-2}$, with $x \in \left[1, 10^2\right]$. As a result, for Bi_2Se_3 one gets

$$\eta \simeq 26\, x^{-1/5}\ , \qquad v_0 \simeq 1.6 \times 10^{-3} x^{4/5}\ , \tag{5.1.25}$$

whereas for IV-VI materials one has

$$\eta \simeq 11\, x^{-1/5}\ , \qquad v_0 \simeq 1.2 \times 10^{-2} x^{4/5}\ . \tag{5.1.26}$$

Although the values are different in both materials, we are not so much interested in obtaining quantitative values, since these are already accurately predicted by the more elaborate methods discussed in [1, 3–6], but rather a more qualitative picture. Hence, we will consider a compromise hereafter and set

$$\eta \simeq 15x^{-1/5}\ , \qquad v_0 \simeq 10^{-2} x^{4/5}\ . \tag{5.1.27}$$

[2]Notice that, in truth, Δ, m^* and v_F are not independent, but are related via $\Delta = m^* v_F^2$, as can be deduced from the bulk dispersion.

5.2 Surface States in δ-doped Boundaries

In this section, we plan to solve the problem of a topological boundary with a δ-layer right at the interface, in order to explore the physics described in the introduction. Therefore, we are facing the following problem

$$\left[-\mathrm{i}\,\alpha_z\partial_\xi + \boldsymbol{\alpha}_\perp\cdot\boldsymbol{\kappa} + \beta\chi(\xi) + v(\xi) - \varepsilon\right]\boldsymbol{\Psi}(\xi) = 0\ , \tag{5.2.1}$$

where we have dropped the subscript in v_{app}. The function $\chi(\xi)$ will be taken to be such that we can describe both a topological boundary, particularizing $\chi(\xi) = \mathrm{sgn}\,(\xi)$, and a system without inversion, in which case $\chi(\xi) = 1$. Rotational symmetry in the XY-plane allows us to introduce cylindrical coordinates, so that we can write

$$\boldsymbol{\kappa}_\perp = \kappa\,(\cos\theta, \sin\theta, 0)\ , \tag{5.2.2}$$

and, as a result,

$$\boldsymbol{\alpha}_\perp\cdot\boldsymbol{\kappa}_\perp = \kappa\,\Sigma\ , \qquad \Sigma = \tau_x\otimes\sigma_\rho\ , \tag{5.2.3}$$

where σ_ρ is the radial component of $\boldsymbol{\sigma}$ in cylindrical coordinates. That is,

$$\sigma_\rho = \sigma_x\cos\theta + \sigma_y\sin\theta\ . \tag{5.2.4}$$

In order to solve this problem, it will prove useful to abandon the orbital-spin basis, $\left\{\psi_c^\uparrow, \psi_c^\downarrow, \psi_v^\uparrow, \psi_v^\downarrow\right\}$. Rather, we shall form linear combinations that mix the orbital degree of freedom but do not mix the spins

$$\boldsymbol{\Phi} = U\,\boldsymbol{\Psi} = \frac{1}{\sqrt{2}}\begin{pmatrix}\psi_v^\uparrow + \psi_c^\uparrow\\ \psi_v^\downarrow - \psi_c^\downarrow\\ \psi_v^\uparrow - \psi_c^\uparrow\\ \psi_v^\downarrow + \psi_c^\downarrow\end{pmatrix}. \tag{5.2.5}$$

We can see that the unitary transformation U is given by

$$U = \frac{1}{\sqrt{2}}\begin{pmatrix}1 & 0 & 1 & 0\\ 0 & -1 & 0 & 1\\ -1 & 0 & 1 & 0\\ 0 & 1 & 0 & 1\end{pmatrix}. \tag{5.2.6}$$

This mixture no longer allows to separate the orbital and spin degrees of freedom and it does not make sense to differentiate between τ_i and σ_i matrices and we shall use simply σ_i. However, we do see that the first and third components still correspond to $\uparrow$, whereas the second and fourth correspond to $\downarrow$. This implies that it is interesting to write

$$\boldsymbol{\Phi} = \begin{pmatrix} \boldsymbol{\phi}_u \\ \boldsymbol{\phi}_l \end{pmatrix}, \qquad \boldsymbol{\phi}_\alpha = \begin{pmatrix} \phi_\alpha^\uparrow \\ \phi_\alpha^\downarrow \end{pmatrix}, \qquad \alpha = u, l \,. \tag{5.2.7}$$

However, careful must be taken with this notation, since it seems to imply that one can write $\boldsymbol{\phi}$ as a product state $\boldsymbol{\phi} = (\phi_u, \phi_l)^T \otimes (\uparrow, \downarrow)^T$, which is not possible since in this basis the orbital and spin degrees of freedom are entangled. That is, it is not possible to write $\phi_\alpha^s = \phi_\alpha \otimes s$ with $s = \uparrow / \downarrow$.

Upon transforming the Hamiltonian, we can write Eq. (5.2.1) as

$$\left[-\mathrm{i}\,\sigma_z \otimes \sigma_0 \partial_\xi + \mathrm{i}\,\kappa \sigma_y \otimes \sigma_z \sigma_\rho - \sigma_x \otimes \sigma_0 \chi(\xi) + v(\xi) - \varepsilon\right] \boldsymbol{\Phi}(\xi) = 0 \,. \tag{5.2.8}$$

Taking into account Eq. (5.2.7), we obtain the following two coupled equations

$$\left[-\mathrm{i}\,\partial_\xi + v(\xi) - \varepsilon\right] \boldsymbol{\phi}_u(\xi) = \left[\chi(\xi) - \kappa \sigma_z \sigma_\rho\right] \boldsymbol{\phi}_l(\xi) \tag{5.2.9a}$$
$$\left[\mathrm{i}\,\partial_\xi + v(\xi) - \varepsilon\right] \boldsymbol{\phi}_l(\xi) = \left[\chi(\xi) + \kappa \sigma_z \sigma_\rho\right] \boldsymbol{\phi}_u(\xi) \,. \tag{5.2.9b}$$

We will solve on both sides of $\xi = 0$ and then apply continuity at the boundary. From Eq. (5.2.9a) we have

$$\boldsymbol{\phi}_l(\xi) = \frac{1}{1 + k^2}\left[\chi(\xi) + \kappa \sigma_z \sigma_\rho\right]\left[-\mathrm{i}\,\partial_\xi + v(\xi) - \varepsilon\right] \boldsymbol{\phi}_u(\xi) \,, \tag{5.2.10}$$

where we have taken into account that $\chi^2(\xi) = 1$. We can then introduce $\boldsymbol{\phi}_l(\xi)$ into Eq. (5.2.9b) to obtain

$$\left[\partial_\xi^2 - \lambda^2 + \mathrm{i}\,\partial_\xi v(\xi) + v^2(\xi) - 2\varepsilon v(\xi)\right] \boldsymbol{\phi}_u(\xi) = 0 \,, \tag{5.2.11}$$

where we have taken into account that $\chi(\xi)$ is independent of ξ when $\xi \neq 0$ and we have defined

$$\lambda^2 = 1 + \kappa^2 - \varepsilon^2 \,. \tag{5.2.12}$$

Notice that Eq. (5.2.11) corresponds to the one we studied in Chap. 3 particularizing $v(\xi)$ to $f\xi$. The term $\partial_\xi v(\xi)$ corresponds to the electric field created by $v(\xi)$. Notice that solving for $\xi > 0$, one can then obtain the solution for $\xi < 0$ by doing $\xi \to -\xi$ and taking the complex conjugate. Solutions for $\xi > 0$ are given by [18, 19]

$$\boldsymbol{\phi}_u(\xi) = \exp\left[-\lambda\xi - \mathrm{i}\,v_0 \eta e^{-\xi/\eta}\right]\left[\boldsymbol{A}_+ \varphi_M(\xi) + \boldsymbol{B}_+ \varphi_U(\xi)\right] \,, \tag{5.2.13}$$

where $\boldsymbol{A}_+$ and $\boldsymbol{B}_+$ are two constant vectors and $\varphi_M(\xi)$, $\varphi_U(\xi)$ are given in terms of Kummer's functions $M(a, b, z)$ and $U(a, b, z)$ [17]

$$\varphi_M(\xi) = M\left(\lambda\eta + \mathrm{i}\,\varepsilon\eta, 1 + 2\lambda\eta, \mathrm{i}\,2v_0\eta e^{-\xi/\eta}\right) \,, \tag{5.2.14a}$$
$$\varphi_U(\xi) = U\left(\lambda\eta + \mathrm{i}\,\varepsilon\eta, 1 + 2\lambda\eta, \mathrm{i}\,2v_0\eta e^{-\xi/\eta}\right) \,. \tag{5.2.14b}$$

Since we are interested in normalizable solutions and $U(a, b, z)$ diverges as $z \to 0$ (i.e. $\xi \to \infty$), then we set $\boldsymbol{B}_+ = 0$. Hence, taking into account the previous comment regarding solutions for $\xi < 0$, we finally find that

$$\boldsymbol{\phi}_u(\xi) = \Theta(\xi)\boldsymbol{A}_+ h(\xi) + \Theta(-\xi)\boldsymbol{A}_- h^*(-\xi) \,, \tag{5.2.15}$$

where $\Theta(\xi)$ is the Heaviside step function and

$$h(\xi) = \exp\left[-\lambda\xi - \mathrm{i}\, v_0 \eta e^{-\xi/\eta}\right] \varphi_M(\xi) \,. \tag{5.2.16}$$

From Eq. (5.2.10) we can find $\boldsymbol{\phi}_l$. However, let us first impose continuity to relate $\boldsymbol{A}_+$ and $\boldsymbol{A}_-$. Since the unitary transformation preserves continuity, we require $\boldsymbol{\Phi}(0^-) = \boldsymbol{\Phi}(0^+)$, which implies that

$$\boldsymbol{A}_- = e^{-\mathrm{i}\theta} \boldsymbol{A}_+ \,, \tag{5.2.17}$$

with

$$\theta = 2v_0\eta - 2\arg\left[M\left(\lambda\eta + \mathrm{i}\,\varepsilon\eta, 1 + 2\lambda\eta, \mathrm{i}\, 2v_0\eta\right)\right] \,. \tag{5.2.18}$$

Therefore, if we define $\boldsymbol{A} \equiv \exp(-\mathrm{i}\theta/2)\boldsymbol{A}_+$, then $\boldsymbol{\phi}_u$ can be written as follows

$$\boldsymbol{\phi}_u = p(\xi) q(\xi) \boldsymbol{A} \,, \tag{5.2.19}$$

where

$$p(\xi) = \Theta(\xi) h(\xi) + \Theta(-\xi) h^*(-\xi) \,, \tag{5.2.20a}$$

$$q(\xi) = \Theta(\xi) e^{\mathrm{i}\theta/2} + \Theta(-\xi) e^{-\mathrm{i}\theta/2} \,. \tag{5.2.20b}$$

Even though this looks like an unnecessarily complicated way of writing $\boldsymbol{\Phi}_u(\xi)$, we shall see why it is interesting to write it like so shortly. Combining this result with Eq. (5.2.10), we obtain $\boldsymbol{\phi}_l(\xi)$

$$\boldsymbol{\phi}_l(\xi) = \frac{\varepsilon - \mathrm{i}\,\lambda s(\xi)}{1 + \kappa^2} p^*(\xi) q(\xi) \left[\chi(\xi) + \kappa\sigma_z\sigma_\rho\right] \boldsymbol{A} \,. \tag{5.2.21}$$

Continuity at the interface implies

$$\left[\mu^*\left(1 + \kappa\sigma_z\sigma_\rho\right) - \mu\left(\nu + \kappa\sigma_z\sigma_\rho\right)\right] \boldsymbol{A} = 0 \,, \tag{5.2.22}$$

where we have defined

$$\mu = (\varepsilon + \mathrm{i}\,\lambda) e^{-\mathrm{i}\theta} \,, \tag{5.2.23}$$

and $\nu = 1$ if there is no inversion, $\nu = -1$ if there is. It is convenient to rewrite this equation as follows

$$\left(\mu - \mu^*\right) \kappa \sigma_z \sigma_\rho \boldsymbol{A} = \left(\mu^* - \mu\nu\right) \boldsymbol{A} \,. \tag{5.2.24}$$

Although this equation is all we need to continue, it is interesting to rearrange this equation a little bit further. On the one hand, we can notice that

$$\kappa \sigma_z \sigma_\rho = -\mathrm{i} \left(\kappa_y \sigma_x - \kappa_x \sigma_y\right) = -\mathrm{i} \left(\boldsymbol{\sigma} \times \boldsymbol{\kappa}\right)_z \,. \tag{5.2.25}$$

On the other hand

$$\mu^* - \mu\nu = \Re[\mu](1-\nu) - \mathrm{i}\,\Im[\mu](1+\nu) \,. \tag{5.2.26}$$

Finally, we can write

$$\Im[\mu] \left(\boldsymbol{\sigma} \times \boldsymbol{\kappa}\right)_z \boldsymbol{A} = \left\{\Re[\mu]\frac{1-\nu}{2} - \mathrm{i}\,\Im[\mu]\frac{1+\nu}{2}\right\} \boldsymbol{A} \,. \tag{5.2.27}$$

We can see that this way of writing Eq. (5.2.22) is already showing signatures of a Rashba effect due to the term on the left-hand side. Let us however explore the two cases of interest separately. Consider that there is no inversion, $\nu = 1$. In this case, we would have

$$\Im[\mu] \left(\boldsymbol{\sigma} \times \boldsymbol{\kappa}\right)_z \boldsymbol{A} = -\mathrm{i}\,\Im[\mu] \boldsymbol{A} \,. \tag{5.2.28}$$

This equation can only be satisfied non-trivially if $\Im[\mu] = 0$. Indeed, if $\Im[\mu] \neq 0$, then this equation would imply that $\boldsymbol{A}$ is an eigenvector of $\left(\boldsymbol{\sigma} \times \boldsymbol{\kappa}\right)_z$ with eigenvalue $-\mathrm{i}$. However, the eigenvalues of $\left(\boldsymbol{\sigma} \times \boldsymbol{\kappa}\right)_z$ are real and, therefore, $\Im[\mu]$ must vanish identically. Hence, there will be a two-fold degeneracy and we can choose $\boldsymbol{A} \sim (1, 0)^T$ and $\boldsymbol{A} \sim (0, 1)^T$ to correspond to each degenerate eigenvalue. The use of $\sim$ indicates that the state must still be normalized. Since the upper (lower) component of $\boldsymbol{A}$ only contains information about spin up (down), we may say that there will be two spin-degenerate solutions. Taking into account the definition of μ, we can write an equation for the energies

$$\lambda \cos\theta - \varepsilon \sin\theta = 0 \,. \tag{5.2.29}$$

We will later show the resulting energies from numerically solving this equation. However, it is interesting to have a comparison with the inverted system, so we first proceed to discuss that situation. In this case, $\nu = -1$ and Eq. (5.2.27) can be written as

$$\Im[\mu] \left(\boldsymbol{\sigma} \times \boldsymbol{\kappa}\right)_z \boldsymbol{A} = \Re[\mu] \boldsymbol{A} \,. \tag{5.2.30}$$

Notice that we cannot have $\Im[\mu] = 0$ since that would require $\Re[\mu] = 0$ for non-trivial solutions and, therefore, $\mu = 0$, which can never be zero as is evident from its definition in Eq. (5.2.23). Hence we may write the previous equation as

$$(\boldsymbol{\sigma} \times \boldsymbol{\kappa})_z \boldsymbol{A} = \frac{\Re[\mu]}{\Im[\mu]} \boldsymbol{A} . \tag{5.2.31}$$

We have finally found what we were looking for. Indeed, we have a Rashba Hamiltonian and, as such, there is a degeneracy breaking, except at $\boldsymbol{\kappa} = 0$. Indeed, at $\boldsymbol{\kappa} = 0$ we must have $\mathrm{Re}[\mu] = 0$ doubly degenerate with $\boldsymbol{A} \sim (1, 0)^T$ and $\boldsymbol{A} \sim (0, 1)^T$. This is compliant with the fact that time-reversal symmetry imposes Kramers' degeneracy at $\boldsymbol{\kappa} = 0$ and the δ layer preserves time-reversal symmetry. Away from $\boldsymbol{\kappa} = 0$, the degeneracy is broken. Before we continue, let us briefly recapitulate what we learnt in Chap. 2. For that matter, consider

$$(\boldsymbol{\sigma} \times \boldsymbol{\kappa})_z \boldsymbol{u} = \Lambda \boldsymbol{u} , \tag{5.2.32}$$

where $\Lambda = \varepsilon$ for the topological surface states, but we shall leave it as a generic Λ since it is of our interest right now. Since there is rotational symmetry, let us pick any $\boldsymbol{\kappa}$ to our convenience, say $\boldsymbol{\kappa} = \kappa_y\, \widehat{\boldsymbol{y}}$. Then, the previous equation is written as

$$\kappa_y \sigma_x \boldsymbol{u} = \Lambda \boldsymbol{u} . \tag{5.2.33}$$

Let us denote the two eigenvectors of σ_x with eigenvalues ± 1 as $|+\rangle_x$ and $|-\rangle_x$. Then, if $\kappa_y = 0$, the two eigenvectors are doubly degenerate with $\Lambda = 0$. On the other hand, if $\kappa_y \neq 0$, then we have $\Lambda_\pm = \pm\kappa_y$ with corresponding eigenvectors $|\pm\rangle_x$. Since these are eigenvectors of σ_x, it is clear that $\langle \boldsymbol{\sigma} \rangle_\pm = \pm\, \widehat{\boldsymbol{x}}$. Hence, we can see that $\langle \boldsymbol{\sigma} \rangle_\pm$ is perpendicular to $\boldsymbol{\kappa}$ and that we have opposite $\langle \boldsymbol{\sigma} \rangle_\pm$ for each $\Lambda_\pm$, that is, $\langle \boldsymbol{\sigma} \rangle_+ = -\langle \boldsymbol{\sigma} \rangle_-$. If we recall for a moment the case of topological surface states in absence of perturbations, where $\Lambda = \varepsilon$, we can see that two branches forming the two Dirac cones, $\varepsilon = \pm\kappa_y$, have opposite $\langle \boldsymbol{\sigma} \rangle_\pm$. The discussion, although particularized to $\boldsymbol{\kappa} = \kappa_y\, \widehat{\boldsymbol{y}}$ is generic to any other direction due to rotational symmetry. This is then what leads to the upper and lower cone having opposite helicities when all directions of $\boldsymbol{\kappa}$ are considered.

With this in mind, let us apply it to the case where $\Lambda = \Re[\mu]/\Im[\mu]$. Taking into account the previous comments about rotational symmetry, we shall denote by κ the component of $\boldsymbol{\kappa}$ in a generic direction. That is, κ is not to be confused with $|\boldsymbol{\kappa}|$, since κ can also be negative. Hence, following the same procedure as before, we find

$$\frac{\Re[\mu]}{\Im[\mu]} = \pm\kappa , \tag{5.2.34}$$

where the $\pm$ signs are associated to $\langle \boldsymbol{\sigma} \rangle_\pm$. The latter, as explained above, is perpendicular to $\boldsymbol{\kappa}$ and $\langle \boldsymbol{\sigma} \rangle_+ = -\langle \boldsymbol{\sigma} \rangle_-$. Taking into account the definition of μ, we can write this equation as follows

$$\tan\theta = \frac{\pm\kappa\lambda - \epsilon}{\lambda \pm \epsilon\kappa} . \tag{5.2.35}$$

Since θ and λ depend only on κ^2, one can see that the solutions for the negative sign can be obtained from those of the positive sign by simply changing $\kappa \to -\kappa$.

Now that we have equations for the energies in both the inverted and non-inverted regimes [cf. Eqs. (5.2.29) and (5.2.35)], we can proceed to solve them numerically. The results are shown in Fig. 5.2, where blue colours correspond to the non-inverted situation, orange and green to the inverted one. In panel (a), we show the evolution of the state at $\kappa = 0$ as the density of donors is increased. The dimensionless electric field close to the boundary is in the order of $v_0/\eta \sim 10^{-3}x$, which implies that the field is directly proportional to the number of donors per unit area. As we can observe, as the number of donors increases, so does the electric field and, therefore, the number of continuum levels sucked in by the Yukawa potential increases. The Dirac point also moves downwards in energy and does so almost exactly by the amount $-v_0$ (dashed line). That is, the Dirac point moves in such a way so as to remain right at the middle of the effective gap at the surface. Very similar results where found in Ref. [23] from a much more elaborate tight-binding model of Bi_2Se_3.

On the other hand, in panels (b) and (c) of Fig. 5.2, we can observe the dispersion when $n_S = 5 \times 10^{11}\,\text{cm}^{-2}$. We also depict the probability densities of states at $\kappa = 0$, together with the profile of $v(\xi)$. There are a number of features to observe in these panels. First, as we pointed out before, the Dirac point is robust, as expected from the topological arguments given above. On the other hand, the electric field leads to a reduction in the Fermi velocity, although it is almost imperceptible since the fields are truly small. In Fig. 5.3a the velocity reduction is shown, together with a quadratic fit (solid line) of the form $1 - \zeta x^2$, with ζ a fitting coefficient. This is very much like the behaviour we predicted in Chap. 3 for the uniform electric field, since the field is proportional to x. Second, extended massive Dirac fermions of the bulk become localized by the Yukawa potential. In panel (c), we can also observe a Rashba splitting in the inverted case, as expected from our previous discussion. Green lines correspond to the solutions of Eq. (5.2.35) with positive sign, thereby corresponding to $\langle\sigma\rangle_+$, whereas orange coloured branches correspond to $\langle\sigma\rangle_-$. It is important to make one important remark: the appearance of this Rashba splitting needs two main ingredients. On the one hand, it requires the electric field derived from the Yukawa potential. This is clearly observed in Fig. 5.3b, where we show the splitting between the first Rashba split subbands, $\Delta\kappa$, as doping increases. As it can be observed from a linear fit (solid line), the splitting is directly proportional to x and, therefore, it is directly proportional to the field. This is what we would expect from a Rashba interaction [24]. The second ingredient that is necessary is structural inversion asymmetry [24]. This is the reason why only the topological boundary displays Rashba splitting, even though both scenarios have a built-in electric field. That is, the Rashba splitting observed here is not topological in origin and would appear also if we considered a non-inverted system where the gaps on both sides of the delta layer are different but not necessarily of opposite sign. Mathematically, this can be observed in Eq. (5.2.27). It is only when $\nu = 1$ that we are led to a situation where only doubly degenerate solutions exist. However, if we considered $\nu \neq 1$, then we would break the degeneracy. In fact, the inverted case with $\nu = -1$ is only a particular case that breaks the degeneracy. What is topological is the fact that only

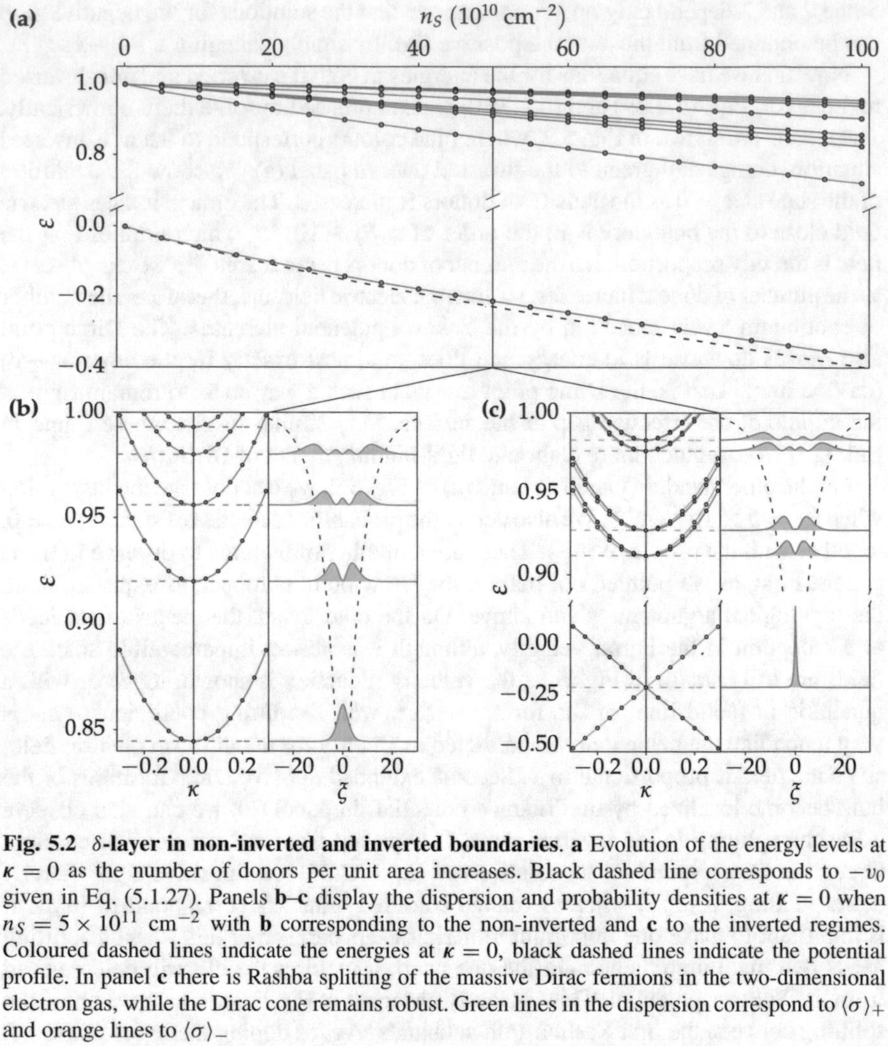

Fig. 5.2 **δ-layer in non-inverted and inverted boundaries.** **a** Evolution of the energy levels at $\kappa = 0$ as the number of donors per unit area increases. Black dashed line corresponds to $-v_0$ given in Eq. (5.1.27). Panels **b–c** display the dispersion and probability densities at $\kappa = 0$ when $n_S = 5 \times 10^{11}$ cm^{-2} with **b** corresponding to the non-inverted and **c** to the inverted regimes. Coloured dashed lines indicate the energies at $\kappa = 0$, black dashed lines indicate the potential profile. In panel **c** there is Rashba splitting of the massive Dirac fermions in the two-dimensional electron gas, while the Dirac cone remains robust. Green lines in the dispersion correspond to $\langle\sigma\rangle_+$ and orange lines to $\langle\sigma\rangle_-$

when sgn $(\nu) = -1$ do we find the topological surface states. Finally, we conclude this section by looking at the probability densities in the non-inverted [cf. panel (b)] and inverted systems [cf. panel (c)]. As we can see, in the absence of band inversion the probability densities resemble those of a quantum well, like the bell-shaped density profile of the lowest state in the well. However, if there is band inversion, the peaked Dirac state prevents the first state of the quantum well to be bell-shaped to ensure orthogonality. Moreover, the density profiles show a non-smooth behaviour right at the boundary as a consequence of band inversion. It is also interesting to observe that the subbands in the inverted regime enter the well in close pairs. This

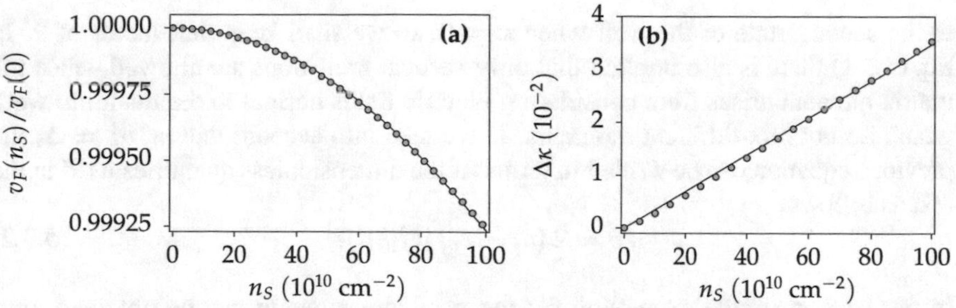

Fig. 5.3 Velocity reduction and Rashba splitting. a The built-in field leads to a slight velocity reduction (the fields are truly small) and, as occurred in previous chapters, this reduction is proportional to the squared field. Solid line is a quadratic fit of the form $1 - \zeta x^2$, with ζ a fitting coefficient. **b** The Rashba splitting, $\Delta\kappa$, is directly proportional to the density of donors, which implies that it is directly proportional to the built-in field. Solid line corresponds to a linear fit to the data

behaviour is truly distinct from the trivial case. Indeed, in the trivial scenario the levels come closer together as a result of the potential widening as we move up in energy.

5.3 Optical Transitions

As an application of the results found in this chapter, we proceed to take advantage of the marked difference between the states in the trivial and topological settings. For that matter, we shall study optical transitions from the first state in the quantum well to the second state. Of course, the Thomas-Fermi approximation is inconsistent with the possibility of observing such transitions, for it assumes that all levels are filled up to the conduction band edge at infinity. In any case, it is interesting to consider how would the oscillator strengths change from one situation to the other. These quantities are key to studying optical absorption [25], since the latter is proportional to the former. Oscillator strengths are independent of the populations and only care about the difference in energies and the shapes of the eigenstates. We expect the states not to change very much if a more complicated model is used instead, so studying the oscillator strengths is interesting in any case. Let $|i\rangle$ be the initial state with energy E_i and $|j\rangle$ the final state with energy E_j. Then, the oscillator strength is defined as [25]

$$f_{ji} = 2m^*(E_j - E_i)|\langle j|z|i\rangle|^2 \,. \tag{5.3.1}$$

The first factor of 2 is included to account for spin degeneracy. However, as we know, Rashba splitting lifts such degeneracy except at the time-reversal invariant momenta. In this case, at $\kappa = 0$. Hence, such a factor of 2 should be removed when studying transitions from momenta different from $\kappa = 0$ if structure inversion asymmetry occurs. We will nevertheless focus exclusively on transitions from the first to the second state of the well when $\kappa = 0$, so we shall keep that factor of 2. In Eq. (5.3.1) there is also implicit that only vertical transitions are allowed, since the matrix element arises from considering electric fields normal to the quantum well, which do not mix different momenta. If we take into account that $m^* v_F^2 = \Delta$, the previous equation can be written in terms of the dimensionless quantities used in the text as follows

$$f_{ji} = 2\left(\varepsilon_j - \varepsilon_i\right) |\langle j|\xi|i\rangle|^2 \,. \tag{5.3.2}$$

In our case, a simple expression for the oscillator strength can be obtained after some manipulations. Let us denote by $\lambda_i = \lambda(\varepsilon_i)$ and $\theta_i = \theta(\varepsilon_i)$, these two quantities being defined in Eqs. (5.2.12) and (5.2.18), respectively. The normalization constants, $N_i = N(\varepsilon_i)$ are given by

$$N_i = \frac{1}{2}\left[\int_0^\infty \mathrm{d}\xi \; |h_i(\xi)|^2\right]^{-1/2} , \tag{5.3.3}$$

where $h_i(\xi) = h(\xi, \varepsilon_i)$ is given in Eq. (5.2.16). If we introduce

$$I_{ji} = \int_0^\infty \mathrm{d}\xi \; \xi \; h_j^*(\xi) h_i(\xi) \,, \tag{5.3.4}$$

then the oscillator strength is simply written as follows

$$f_{ji} = 8\left(\varepsilon_j - \varepsilon_i\right) \left\{\Im[G_{ji}]\right\}^2 \,, \tag{5.3.5}$$

where

$$G_{ji} = N_j N_i I_{ji}\left[e^{-\mathrm{i}\theta_{ji}} - e^{\mathrm{i}\theta_{ji}}\left(\varepsilon_j - \mathrm{i}\,\lambda_j\right)\left(\varepsilon_i + \mathrm{i}\,\lambda_i\right)\right] , \tag{5.3.6}$$

with $\theta_{ji} = (\theta_j - \theta_i)/2$. Using this result, the oscillator strength related to transitions from the first level of the well potential to the second level, f_{21}, as a function of the density of donors is shown in Fig. 5.4. The behaviour of the oscillator strengths is notably different in both cases, which is understandable taking into account the probability density profiles that we discussed previously. As we can see, f_{21} saturates quite rapidly in the non-inverted situation (blue) and is close to one. The oscillator strengths follow a sum rule [25]

$$\sum_{j, j \neq i} f_{ji} = 1 \,. \tag{5.3.7}$$

Then, it is clear that the transition $1 \to 2$ will be much stronger than all other transitions from 1 to other levels. In contrast, f_{21} increases steadily with the doping density in the topological case.

Evidently, as we saw in the third chapter, including a field normal to the interface leads to nontrivial behaviours, which are not accounted for in this simplified

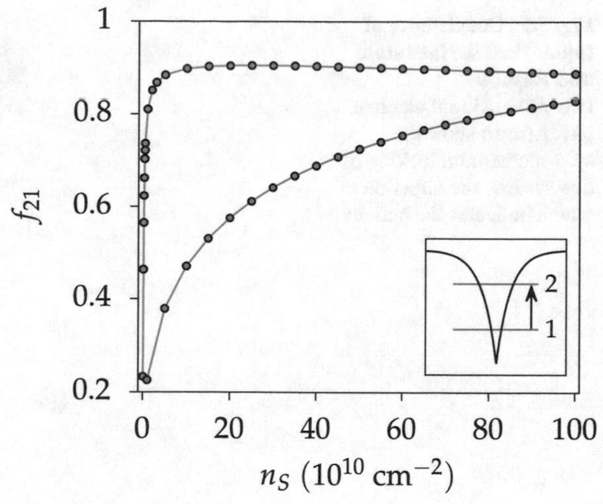

Fig. 5.4 Oscillator strength of intraband transitions from the first to the second level in the Yukawa potential. Blue and orange colours correspond to the non-inverted and inverted scenarios, respectively

description of optical transitions. However, it is interesting to observe the striking difference between the two scenarios, suggesting that optical studies could be conducted in order to unravel the properties of topological insulators.

5.4 Conclusions

In this chapter, we have considered the effect of placing a dense δ layer of donor impurities at a topological boundary. This perturbation does not break time-reversal symmetry, implying the robustness of the Dirac cone. Notice that this is also applicable to topological crystalline insulators, since this perturbation does not break mirror symmetry about those planes perpendicular to the XY-plane. Microscopically, however, mirror symmetry would indeed be broken, although in the high-density limit the possible gap opening at the Dirac point should be negligible. On the other hand, we have observed how the screened Coulomb potential as obtained from Thomas-Fermi localizes states of the continuum forming subbands. If structural inversion asymmetry occurs, Rashba splitting takes place. Interestingly enough, the Dirac state coexists with the Rashba-split two-dimensional electron gas. A schematic depiction is shown (Fig. 5.5).

Finally, the oscillator strength gets reshaped dramatically upon having band inversion with respect to the trivial case. Although the point where f_{21} saturates in the trivial case an the growth rate in the topological case depend on the specific parameters chosen, the behaviour is still very different in any case. Indeed, in Ref. [26], $v_0 \simeq 2 \times 10^{-2} x^{4/5}$ and $\eta \simeq 5.5 x^{-1/5}$ and the results are different quantitatively, but still a similar behaviour was observed.

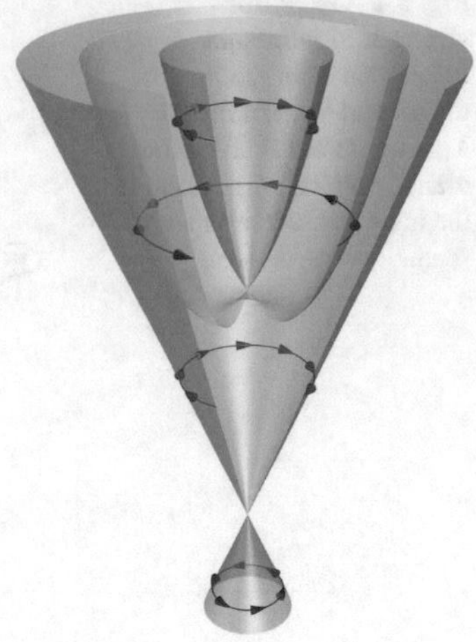

Fig. 5.5 Coexistence of topological surface state and Rashba two-dimensional electron gas. Arrows show the spin-momentum locking of this system, the direction of which indicates the helicity

References

1. Bianchi M, Guan D, Bao S, Mi J, Iversen BB, King PDC, Hofmann P (2010) Coexistence of the topological state and a two-dimensional electron gas on the surface of Bi_2Se_3. Nat Commun 1:128
2. Benia HM, Lin C, Kern K, Ast CR (2011) Reactive chemical doping of the Bi_2Se_3 topological insulator. Phys Rev Lett 107:177602
3. King PDC, Hatch RC, Bianchi M, Ovsyannikov R, Lupulescu C, Landolt G, Slomski B, Dil JH, Guan D, Mi JL, Rienks EDL, Fink J, Lindblad A, Svensson S, Bao S, Balakrishnan G, Iversen BB, Osterwalder J, Eberhardt W, Baumberger F, Hofmann P (2011) Large tunable rashba spin splitting of a two-dimensional electron gas in Bi2Se3. Phys Rev Lett 107:096802
4. Zhu Z-H, Levy G, Ludbrook B, Veenstra CN, Rosen JA, Comin R, Wong D, Dosanjh P, Ubaldini A, Syers P, Butch NP, Paglione J, Elfimov IS, Damascelli A (2011) Rashba spin-splitting control at the surface of the topological insulator Bi2Se3. Phys Rev Lett 107:186405
5. Bianchi M, Hatch RC, Mi J, Iversen BB, Hofmann P (2011) Simultaneous quantization of bulk conduction and valence states through adsorption of nonmagnetic impurities on Bi_2Se_3. Phys Rev Lett 107:086802
6. Bahramy MS, King PDC, de la Torre A, Chang J, Shi M, Patthey L, Balakrishnan G, Hofmann Ph, Arita R, Nagaosa N, Baumberger F (2012) Emergent quantum confinement at topological insulator surfaces. Nat Commun 3:1159
7. Chen C, He S, Weng H, Zhang W, Zhao L, Liu H, Jia X, Mou D, Liu S, He J, Peng Y, Feng Y, Xie Z, Liu G, Dong X, Zhang J, Wang X, Peng Q, Wang Z, Zhang S, Yang F, Chen C, Xu Z, Dai X, Fang Z, Zhou XJ (2012) Robustness of topological order and formation of quantum well states in topological insulators exposed to ambient environment. Proc Natl Acad Sci USA 109:3694
8. Whall TE (1992) A plain man's guide to delta-doped semiconductors. Contemp Phys 33:369
9. Schubert EF (1996) Delta-doping of semiconductors. Cambridge University Press, Cambridge
10. Fermi E (1927) Application of statistical gas methods to electronic systems. Rend Accad Naz Lincei 6:602
11. Thomas LH (1927) Calculation of atomic fields. Proc Camb Philos Soc 33:542
12. Spruch L (1991) Pedagogic notes on Thomas-Fermi theory (and on some improvements): atoms, stars, and the stability of bulk matter. Rev Mod Phys 63:151
13. Giuliani G, Vignale G (2005) Quantum theory of the electron liquid. Cambridge University Press, Cambridge

14. Ioriatti L (1990) Thomas-Fermi theory of δ-doped semiconductor structures: exact analytical results in the high-density limit. Phys Rev B 41:8340
15. Ihn T (2009) Semiconductor nanostructures: quantum states and electronic transport. Oxford University Press
16. Bastard G (1991) Wave mechanics applied to semiconductor heterostructures. Les Editions de Physique, Les Ulis
17. Abramowitz M, Stegun I (1972) Handbook of mathematical functions. Dover, New York
18. Domínguez-Adame F, Rodríguez A (1995) A one-dimensional relativistic screened Coulomb potential. Phys Lett A 198:275
19. Domínguez-Adame F (1996) Subband energy in two-band δ-doped semiconductors. Phys Lett A 211:247
20. Tchoumakov S, Jouffrey V, Inhofer A, Plaçais B, Carpentier D, Goerbig MO (2017) Volkov-Pankratov states in topological heterojunctions. Phys Rev B 96:201302
21. Littlewood PB (1979) The dielectric constant of cubic IV-VI compounds. J Phys C 12:4459
22. Korenman V, Drew HD (1987) Subbands in the gap in inverted-band semiconductor quantum wells. Phys Rev B 35:6446
23. Park K, De Beule C, Partoens B (2013) The ageing effect in topological insulators: evolution of the surface electronic structure of Bi_2Se_3 upon K adsorption. New J Phys 15:113031
24. Winkler R (2003) Spin-Orbit Coupling Effects in Two-Dimensional Electron and hole systems. Springer, Berlin
25. Davies JH (1997) The physics of low-dimensional semiconductors: an introduction. Cambridge University Press, Cambridge
26. Díaz-Fernández A, del Valle N, Díaz E, Domínguez-Adame F (2018) Topologically protected states in delta-doped junctions with band inversion. Phys Rev B 98:085424

Chapter 6
Floquet Engineering of Dirac Cones

In previous chapters, we have explored the physics of topological surface states under external perturbations and doping. However, during the last decades it has become apparent that the use of periodic drivings can lead to a plethora of new possibilities. Indeed, the potential of such techniques has permeated not only the area of condensed matter physics and cold atomic systems [see Ref. [1] for a review], but also acoustics [2] and photonics [3]. In this scenario, one invokes Floquet's theory [4–7]. To the most basic level, Floquet's theory is nothing but Bloch's theory in the time-domain.[1] A central concept of the theory is that of *quasienergies* [6]. Although these may sound exotic, one may recall the quasimomentum in a periodic lattice. Crudely speaking, quasimomentum is equivalent to ordinary momentum in that it is conserved, the difference being that the conservation of the former is only modulo a reciprocal lattice vector. In order to have conservation of total momentum in collisions, the excess momentum is absorbed by the lattice itself. The idea is basically the same with quasienergies and ordinary energy. In a system that is continuously translationally invariant in time, energy is conserved. When such invariance only occurs discretely, then energy is not conserved anymore, but quasienergies are, modulo the driving frequency. In this case, the excess energy is absorbed by the environment in terms of emitted or absorbed photons, thereby leading to conservation of the total energy. Hence, one can introduce Floquet-Brillouin zones and many concepts that arise in time-independent systems are directly transferred to these setups. In particular, concepts of topology discussed in previous chapters also appear. Interestingly, however, periodic drivings allow for extended tunability since the frequency, intensity and polarization of the driving fields can be easily manipulated by external knobs. In contrast, relying on the natural properties of a solid state system leads to limitations that are hard to overcome otherwise.

In this chapter, far from exploring all the possibilities allowed by Floquet's theory, which would require a whole new Thesis, we shall restrict to studying the quasienergy

[1] To be fair, Bloch's theory is nothing but Floquet's theory in the space of space-periodic systems, for Floquet developed his theory in 1883 [4], much earlier than Felix Bloch was even born.

Á. Díaz Fernández, *Reshaping of Dirac Cones in Topological Insulators and Graphene*, Springer Theses, https://doi.org/10.1007/978-3-030-61555-0_6

spectrum of a topological boundary when a periodic driving is applied. The study of topology in Floquet systems has been considered extensively throughout the past years, see e.g. [8–14]. In particular, manipulating Dirac cones in the quasienergy spectrum has received both theoretical [10, 11, 13, 15, 16] and experimental [17] attention in surfaces of topological insulators and graphene. The results show that in-plane circularly polarized light leads to gap openings in the otherwise gapless spectrum. In graphene, it has also been predicted that linear in-plane fields would lead to anisotropic Dirac cones [15]. In our contribution, we shall consider other configurations additional to the ones discussed in the literature. Moreover, the aforementioned references focus on the effective Hamiltonian for the surface states, performing perturbation theory in the high-frequency limit [10]. In our case, we will also consider such a limit, although using the full Hamiltonian of the topological boundary. This allows for the introduction of out-of-plane fields that are not accounted for in the previously mentioned works. Although we shall not discuss it in detail, the usage of the full Hamiltonian allows us to observe the interplay between surface and bulk states, which are not accessible to the effective surface Hamiltonian [18]. Finally, we would like to remark that the author has recently shown that analytical expressions obtained by means of high-frequency expansions up to second order perfectly match the numerical calculations, providing further support to the results presented herein. Moreover, the analytical expressions provide exact dependences on the driving field parameters. However, since these results were derived after presenting this thesis, they shall not be discussed in this chapter and the reader is encouraged to consult Ref. [19].

6.1 Periodically-Driven Topological Boundary

In this section, we shall explore Floquet's theory using the Hamiltonian for a symmetric-centered topological boundary. In order to include the driving field, as we did with the magnetic field in the fourth chapter, we perform the substitution

$$\boldsymbol{p} \to \boldsymbol{p} + \boldsymbol{A}(t)\,, \tag{6.1.1}$$

where we consider $\boldsymbol{A}(t) = \boldsymbol{A}(t+T)$ independent of position, with T the driving period. In other words, the sample's size is sufficiently small so as to consider spatial variations in the driving field to be negligible [16]. Hence, the resulting electric field derived from this potential is simply $\boldsymbol{F}(t) = -\partial_t \boldsymbol{A}(t)$. It is not particularly difficult to show that a unitary transformation links the description upon performing the substitution (6.1.1) to the one considering a potential $V(\boldsymbol{r},t) = \boldsymbol{r}\cdot\boldsymbol{F}(t)$. Nevertheless, solving the problem turns out to be simpler by performing the aforementioned substitution. Hence, the problem we are facing is the following

$$\mathrm{i}\,\partial_t \boldsymbol{\Psi}(\boldsymbol{r},t) = \left[\boldsymbol{\alpha}\cdot(\boldsymbol{p}+\boldsymbol{A}(t)) + \beta\,\mathrm{sgn}\,(z)\right]\boldsymbol{\Psi}(\boldsymbol{r},t)\,. \tag{6.1.2}$$

Taking into account the periodicity of $\boldsymbol{A}(t)$ along with translational symmetry in the XY plane, we ask for solutions of the form

$$\boldsymbol{\Psi}(\boldsymbol{r}, t) = e^{-\mathrm{i}\varepsilon t} e^{\mathrm{i}\boldsymbol{\kappa}\cdot\boldsymbol{r}_\perp} \boldsymbol{\Phi}(z, t) , \qquad \boldsymbol{\Phi}(z, t) = \boldsymbol{\Phi}(z, t + T) . \tag{6.1.3}$$

Here ε is the quasienergy that we talked about in the introduction of this chapter. Notice the similarity with the quasimomentum in a lattice. Indeed, from (6.1.3) it is clear that ε is periodic and can be restricted to be within the first Floquet-Brillouin zone $\varepsilon \in [-\omega/2, \omega/2)$. If we Fourier expand $\boldsymbol{\Phi}(z, t)$ and $\boldsymbol{A}(t)$, then Eq. (6.1.2) can be written as

$$\varepsilon \boldsymbol{\varphi}_l(\xi) = \left[\alpha_z p_z \boldsymbol{\alpha}_\perp \cdot \boldsymbol{\kappa} + \beta \mathrm{sgn}\,(z) - l\omega\right] \boldsymbol{\varphi}_l(z) + \sum_m \left[\boldsymbol{\alpha} \cdot \boldsymbol{A}_m\right] \boldsymbol{\varphi}_{l-m}(z) , \tag{6.1.4}$$

where $\boldsymbol{\varphi}_l(z)$ and $\boldsymbol{A}_m$ are the Fourier components of $\boldsymbol{\Phi}(z, t)$ and $\boldsymbol{A}(t)$, respectively. The index l runs over the integers. As it is, the equation for the Fourier components is generic to any time-dependent driving. The first thing that is already clear is that if we set $\boldsymbol{A}_m = 0$, we obtain an infinite series of identical decoupled time-independent topological boundary Hamiltonians with energies $\varepsilon + l\omega$. Therefore, we would expect to see replicas of the spectrum of the topological boundary. If we considered the first Floquet-Brillouin zone, we would see evenly spaced Dirac cones where the spacing between Dirac points is ω. This result is identical to that of artificially folding the parabolic band structure of a free electron gas into a first Brillouin zone. On the other hand, whenever $\boldsymbol{A}_m \neq 0$, then different Fourier components become coupled depending on the form of the perturbation. The result that we would expect would be avoided crossings right at the Floquet-Brillouin zone edges [20]. Once again, if we considered the free electron gas picture, when we turn on the crystal potential, at the edges of the Brillouin zone gaps are likely to appear. In our case, it is clear from the form of the perturbation that this must occur since one cannot choose $\boldsymbol{\varphi}_l(z)$ such that the Fourier components become decoupled. In order to make further progress, let us particularize the driving to be

$$\boldsymbol{A}(t) = \boldsymbol{a} e^{\mathrm{i}\,\omega t} + \boldsymbol{a}^* e^{-\mathrm{i}\,\omega t} , \qquad a_j = \frac{f_j}{2\omega} e^{\mathrm{i}\theta_j} , \tag{6.1.5}$$

where f_j is the field amplitude along the j-th direction and θ_j is a phase. By choosing different amplitudes and phases one can explore different polarizations and field orientations. This form of the vector potential is particularly appealing since it is such that only nearest-neighbouring Fourier components get mixed by it. Indeed, in this case only $\boldsymbol{A}_1$ and $\boldsymbol{A}_{-1}$ are nonzero and Eq. (6.1.4) can be written as

$$\varepsilon \boldsymbol{\varphi}_l(z) = \mathcal{H}_l^0 \boldsymbol{\varphi}_l(z) + J \boldsymbol{\varphi}_{l+1}(z) + J^\dagger \boldsymbol{\varphi}_{l-1}(z) , \tag{6.1.6}$$

with $J = \boldsymbol{\alpha} \cdot \boldsymbol{a}$ and

$$\mathcal{H}_l^0 = \alpha_z p_z + \boldsymbol{\alpha}_\perp \cdot \boldsymbol{\kappa} + \beta \mathrm{sgn}\,(z) - l\omega\,. \tag{6.1.7}$$

As it can be observed, if there was no boundary, then Eq. (6.1.6) would correspond to a nearest-neighbour tight-binding model in a one-dimensional chain with four degrees of freedom per site. It is then clear that the Floquet formalism is introducing an extra synthetic dimension to the driving-free problem. From this equation, one can already make predictions on the robustness of the Dirac point. Indeed, let us consider a linearly polarized field. In such a case, it is possible to choose the phases to be zero, resulting in real-valued $\boldsymbol{a}$'s and, therefore, Hermitian hopping matrices J. Alternatively, one can write $J = \tilde{J}\exp(\mathrm{i}\,\theta)$ with $\tilde{J}$ being Hermitian and remove the phase factors via a gauge transformation of the form $\boldsymbol{\varphi}_l \to \exp\left[-\mathrm{i}\,(l-1)\theta\right]\boldsymbol{\varphi}_l$. This in turn implies that time-reversal symmetry is not broken in this situation and we expect the Dirac point to be robust. On the other hand, if the polarization is circular, these manipulations can no longer be done and, as a result, J is non-Hermitian in general, thereby breaking time-reversal symmetry. However, if the field is out-of-plane, then the projection onto the topological boundary is linearly polarized. Since the Dirac state lives at the surface, it is expected that such a situation would preserve the Dirac point as well.

Equation (6.1.6) has to be solved numerically. The details of the numerical implementation are left as an appendix. The idea is to place the system in a box of size $L > 1$ in the Z-direction and discretize the real space variable in a one-dimensional lattice. Instead of sampling all four components of the bispinor in every lattice site, we follow Ref. [21] by sampling the components of the bispinor in an alternating fashion. In particular, we will sample the first and last components of the bispinor on the even sites and the middle components on the odd sites. At the same time, one imposes a cutoff to the sideband or Fourier index. In dealing with this problem, one has to take great care to separate the bulk to the surface physics. Indeed, since we have placed the system in a box, the bands in the continuum will form subbands and these will enter the first Floquet-Brillouin zone upon band folding. In order to see whether the Dirac state remains localized at the boundary despite the application of the external field, one can optimize the size of the box and the discretization step. Indeed, if the box increases in size or the discretization step decreases in size, there will be more bulk quasienergies within the first Floquet-Brillouin zone. However, if upon doing so the Dirac state remains unaltered, then we can conclude that it is localized at the boundary and it is well separated from the bulk states so that there is no hybridization. We shall also consider the high-frequency limit. This limit implies that we choose the frequencies to be larger than all other energy scales in the problem, in this case the energy gap. That is, we assume the dynamics of the system to occur in time scales much larger than the driving period, so that we obtain a quasi-static behaviour. Hence, we set $\omega > 2$. Additional to this requirement, we ask for the driving amplitudes to be small, in such a way that $f/\omega < 1$ so that the perturbations a_j are also small. These requirements are such that one can observe similar physics as with the driving-free system, such as the reshaping of the Dirac cones, with added value from the different polarizations. In what follows, we will consider in-plane and out-of-plane fields, as well as linear and circular polarizations. The results will be discussed within each subsection.

6.1.1 In-Plane Fields

In this subsection, we consider in-plane fields ($f_z = 0$) with linear and circular polarizations. As said previously, in order to assess the localization of the surface state, we will solve for $L = 3$ and two different lattice constants of 0.300 and 0.375. Additionally, we set a cutoff to the sideband index at $l = 3$. In both cases, we fix $\omega = 4$. In the linearly polarized case, we set $f_x = 2$ and $f_y = 0$, and in the circularly polarized case we set $f_x = f_y = 2$ and $\theta_x = \pi/2$, all other phases equal to zero. The quasienergy spectrum is shown in Fig. 6.1.

There are a number of features to observe in this figure. First, upon decreasing the discretization step, the number of bulk subbands increases, as expected. However, the Dirac state is unchanged upon changing the step and the dispersions overlap. Next, we can observe that there are avoided crossings at the edges of the Floquet-Brillouin zone, except for the Dirac state in Fig. 6.1 (a). Following Ref. [20], this can be understood from the fact that the perturbation $f_x\alpha_x$ commutes with $\boldsymbol{\alpha}_\perp \cdot \boldsymbol{\kappa}$ when $\kappa_y = 0$, whereas it does not when $\kappa_x = 0$. Hence, the perturbation does not couple the Dirac sidebands in the first case. Another observation that can be made is the fact that, due to the need to perform avoided crossings at the edges of the Floquet-Brillouin zone, the slope of the Dirac spectrum is reduced for low momenta, thereby revealing the same physics as in previous chapters. Hence, the dispersion is an anisotropic cone, widening in the direction perpendicular to the perturbation, very much like in the case of an in-plane magnetic field [cf. Chap. 4]. This result is similar to what has been found for graphene in Ref. [15]. In our case, however, we are proving that this also occurs in topological insulators, despite the presence of bulk states. Hence, our results confirm that one may still utilize an effective surface Hamiltonian to model the physics discussed here, since the bulk states and the surface states remain uncoupled. In order to keep the same notation as in previous chapters, we shall denote the slope of the cone by $v_F(f)$ and will be called the Fermi velocity hereafter. In the right-hand panel of Fig. 6.2, a schematic depiction of the anisotropic widening is shown. Additionally, the reduction of the Fermi velocity as a function of f/ω for three values of ω is also shown in that figure, together with a quadratic fit of the form $1 - \zeta(f/\omega)^2$, with ζ a fitting coefficient. As it can be observed, the fit is noteworthy and shows that the behaviour explored herein is very similar to that discussed in the static case. Here we have set $L = 5$, $a = 0.5$ and $N_\omega = 3$ and the Fermi velocity is obtained from fitting the linear dispersion up to $\kappa = 0.5$.

In the case of circular polarization [cf. Fig. 6.1c, d], we can see that the dispersion is isotropic and an energy gap, 2δ, opens up at the Dirac point, as expected from breaking time-reversal symmetry. As it can be observed, there are no avoided crossings with the states in the bulk, meaning that these remain essentially uncoupled from the massive surface states. Once again, our results confirm the validity of using an effective

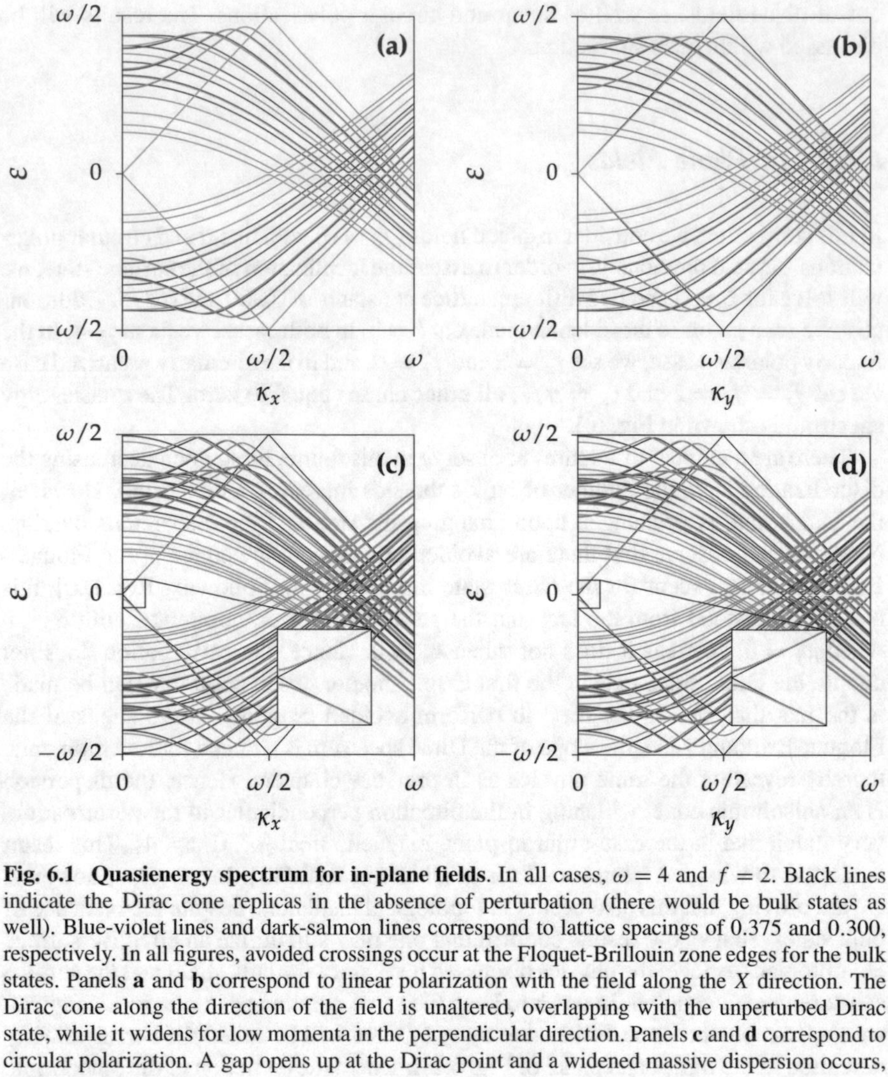

Fig. 6.1 Quasienergy spectrum for in-plane fields. In all cases, $\omega = 4$ and $f = 2$. Black lines indicate the Dirac cone replicas in the absence of perturbation (there would be bulk states as well). Blue-violet lines and dark-salmon lines correspond to lattice spacings of 0.375 and 0.300, respectively. In all figures, avoided crossings occur at the Floquet-Brillouin zone edges for the bulk states. Panels **a** and **b** correspond to linear polarization with the field along the X direction. The Dirac cone along the direction of the field is unaltered, overlapping with the unperturbed Dirac state, while it widens for low momenta in the perpendicular direction. Panels **c** and **d** correspond to circular polarization. A gap opens up at the Dirac point and a widened massive dispersion occurs, as observed in the inset

surface Hamiltonian to obtain results about the physics of the irradiated surface. It is interesting to observe that, additionally to the gap opening, there is a reduction in the slope as well. In fact, both quantities change with f/ω in a quadratic fashion, as shown in Fig. 6.4. The system size, spacing and sideband index cutoff are the same as before, the gap and the Fermi velocity being obtained from fitting the dispersion to a massive Dirac spectrum up to $\kappa = 0.5$. Finally, before we move on to the next section, it is interesting to observe how the energy gap changes as a function of polarization. Indeed, if we denote by δ_{xy} the dephasing between the X and Y components of the

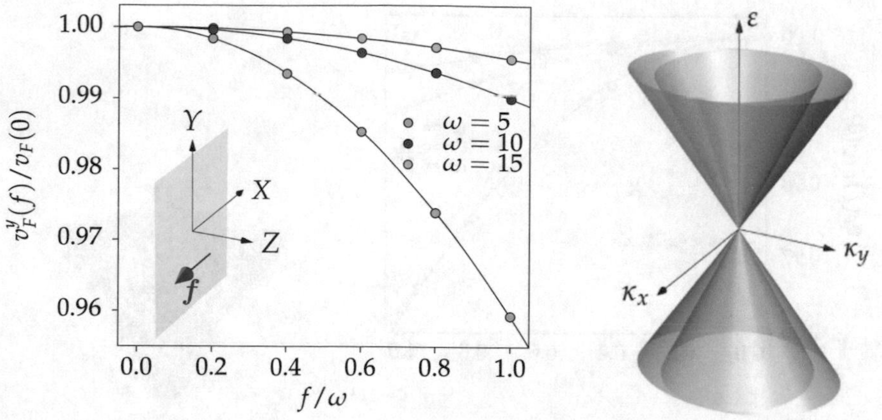

Fig. 6.2 Anisotropic Dirac cones. The field is contained within the topological boundary and it is linearly polarized. As observed in the schematic depiction, the original cone (red) widens in the direction perpendicular to the field to form an anisotropic cone (blue). The panel on the left shows the velocity reduction for different driving frequencies, together with quadratic fits (solid lines) of the form $1 - \zeta(f/\omega)^2$, with ζ a fitting parameter

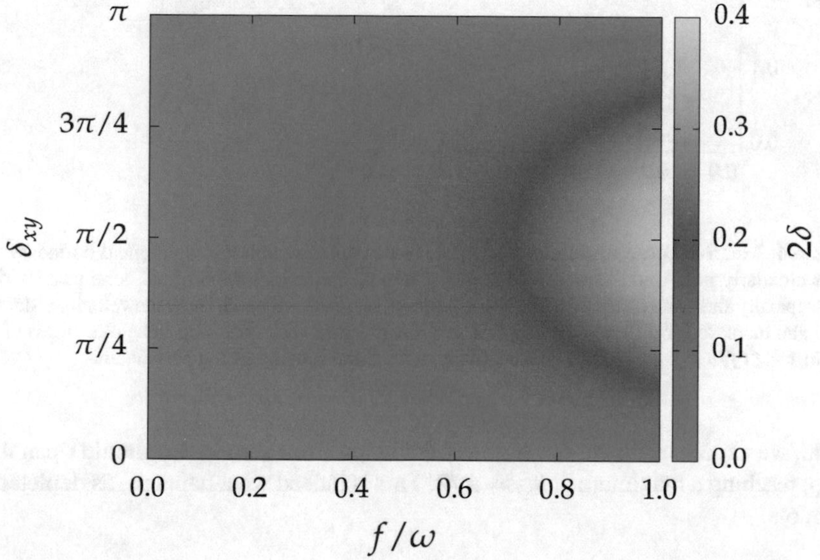

Fig. 6.3 Gap as a function of polarization. When the dephasing is an integer multiple of π, time-reversal symmetry is unbroken and the gap closes up. However, any dephasing within the region $(0, \pi)$ leads to gap openings, which reach a maximum at $\delta_{xy} = \pi/2$ (circular polarization) and increase upon increasing the field strength. In this figure $\omega = 5$

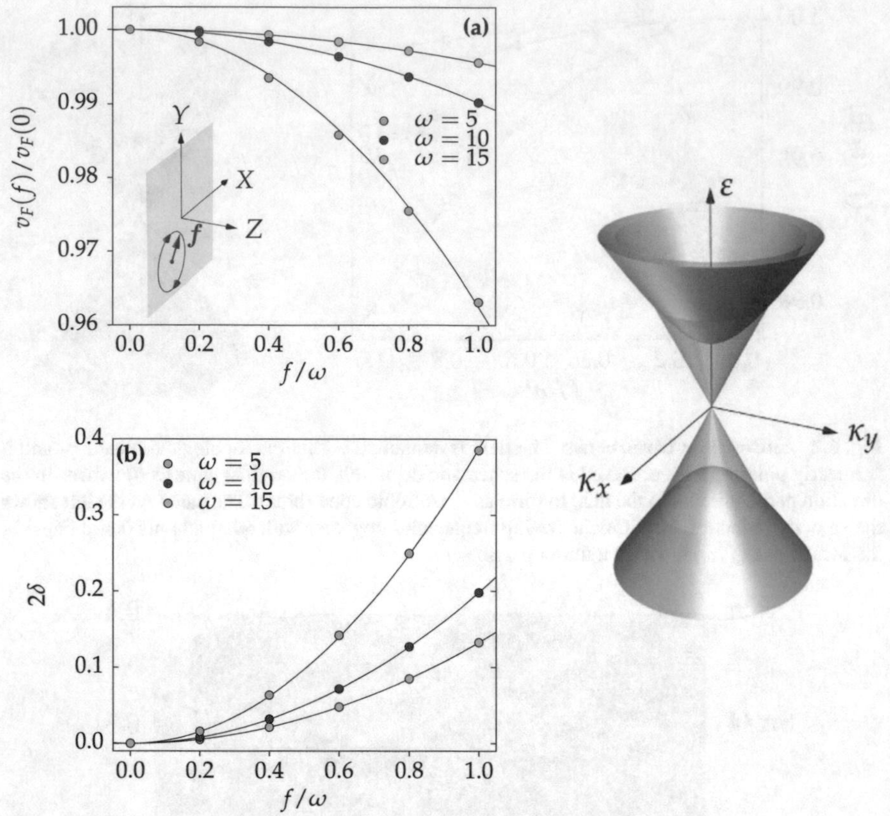

Fig. 6.4 Massive Dirac dispersion. The field is contained within the topological boundary and it is circularly polarized. As observed in the schematic depiction, the original cone (red) widens isotropically and a gap opens up at the Dirac point (blue). Panels **a** and **b** show the velocity reduction and gap increase for different driving frequencies, together with quadratic fits (solid lines) of the form $1 - \zeta(f/\omega)^2$, and $\lambda(f/\omega)^2$, respectively, with ζ and λ being fitting parameters

field, we expect that even the slightest deviation from zero in δ_{xy} should open up a gap, reaching a maximum at $\delta_{xy} = \pi/2$. This is indeed what happens, as depicted in Fig. 6.3.

6.1.2 Out-of-Plane Fields

In this section, we consider the same settings, i.e. system size, lattice parameters and frequency, as in the previous one, except that the fields will be out-of-plane. In all cases, $f_x = 0$. We set $f_y = 0$ and $f_z = 2$ for the linearly polarized case and $f_y = f_z = 2$ and $\theta_x = \pi/2$ for the circularly polarized case, all other phases equal

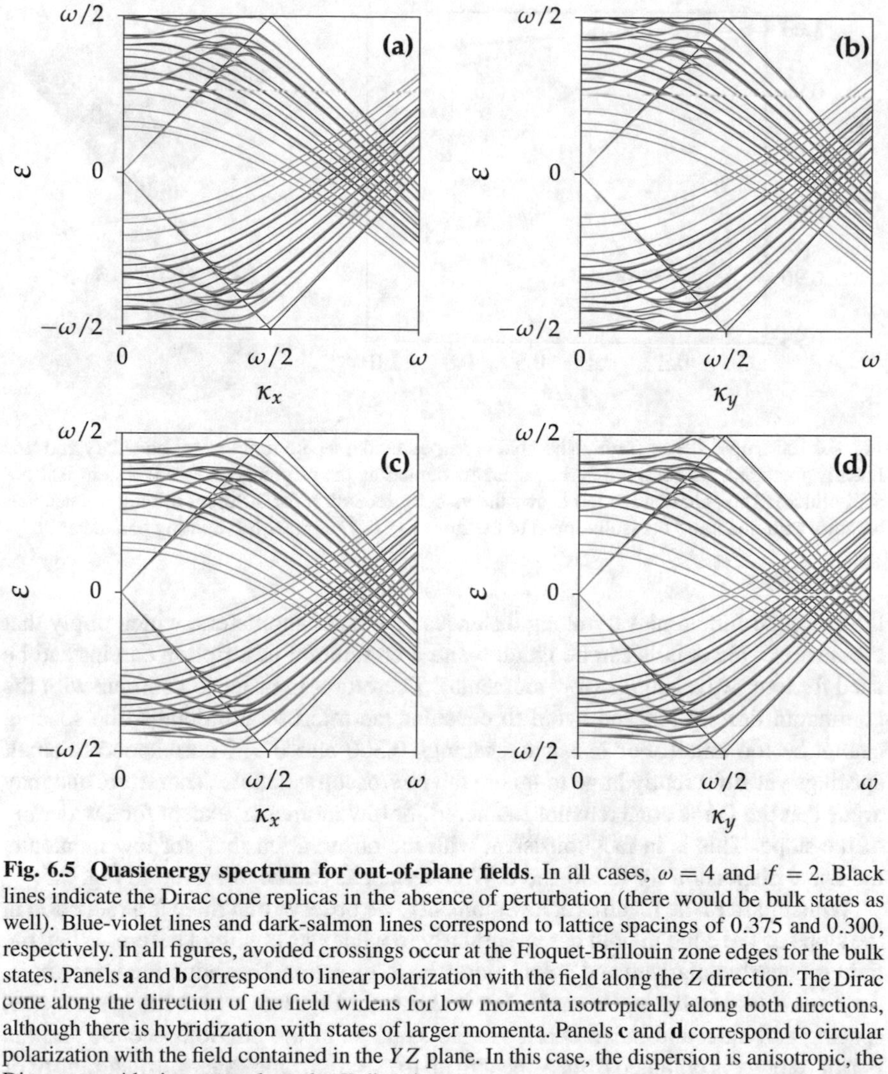

Fig. 6.5 Quasienergy spectrum for out-of-plane fields. In all cases, $\omega = 4$ and $f = 2$. Black lines indicate the Dirac cone replicas in the absence of perturbation (there would be bulk states as well). Blue-violet lines and dark-salmon lines correspond to lattice spacings of 0.375 and 0.300, respectively. In all figures, avoided crossings occur at the Floquet-Brillouin zone edges for the bulk states. Panels **a** and **b** correspond to linear polarization with the field along the Z direction. The Dirac cone along the direction of the field widens for low momenta isotropically along both directions, although there is hybridization with states of larger momenta. Panels **c** and **d** correspond to circular polarization with the field contained in the YZ plane. In this case, the dispersion is anisotropic, the Dirac cone widening more along the X direction

to zero. In this case, we expect the Dirac point to be robust since there is no time-reversal symmetry breaking. However, as shown in Fig. 6.5, there is hybridization with states in the bulk for large momenta. This can be understood by appealing to the static case, where hybridization is more likely to occur closer to the band edges due to proximity to the bulk states, as we discussed in Chap. 3. As the number of bulk states increases due to decreasing of the lattice spacing, the avoided crossings with the Dirac state occur closer to the Dirac point. One may argue that, however, it is not possible to reduce the lattice spacing as much as one would like. Indeed,

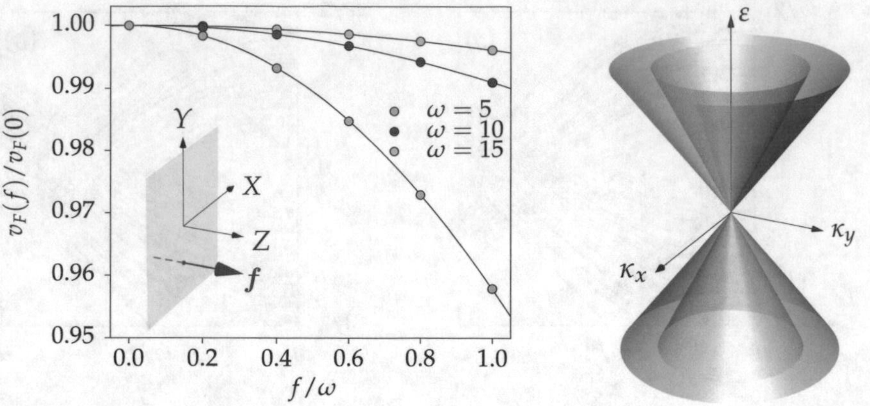

Fig. 6.6 **Isotropic Dirac cones**. The field is perpendicular to the topological boundary and it is linearly polarized. As observed in the schematic depiction, the original cone (red) widens isotropically (blue). The panel on the left shows the velocity reduction for different driving frequencies, together with quadratic fits (solid lines) of the form $1 - \zeta(f/\omega)^2$, with ζ a fitting parameter

the envelope functions vary along distances of a few nanometers, which imply that the microscopic details can be ignored and a continuum description can instead be used for long wavelengths (low momenta). Therefore, in order to continue with the continuum description and avoid to consider the microscopic details, the spacing cannot be too small. For instance, spacings 0.300 and 0.375 correspond to small spacings yet sufficiently large to ignore the microscopic details. Therefore, one may argue that the Dirac cone remains unaltered for low momenta, except for a widening of the slope. This is in fact consistent with the observation that, for low momenta, the Dirac dispersion for 0.300 and 0.375 overlap, as can be observed in Fig. 6.5.

With regard to the features at low momenta, we observe that there is a reduction in the slope of the cone in both cases, similarly to what was obtained before. When the field is linearly polarized and points along the Z-direction, the reduction is isotropic, very much like in the static case. This is shown in Fig. 6.6, where the system size, spacing and sideband index cutoff are the same as in the previous section and the Fermi velocity is obtained from a linear fit up to $\kappa = 0.5$. Once again, a quadratic fit of the form $1 - \zeta(f/\omega)^2$, with ζ a fitting parameter, provides perfect agreement with the data. Similarly, when the field is circularly polarized and contained in the YZ-plane, the cone widens anisotropically, with increased widening along the X-direction. This can be understood by decomposing the circularly polarized field into a linearly polarized component along the Y-direction and another along the Z-direction. As we know, the first component will lead to a reduction only along the X-direction. In contrast, the second component will do so isotropically in both directions. The net result is an enhanced reduction along the X-direction with respect to that along the Y-direction. This is shown in Fig. 6.7, together with fits of the form $1 - \zeta(f/\omega)^2$, with ζ a fitting parameter. The Fermi velocity is obtained from a linear fit of the dispersion up to $\kappa = 0.5$.

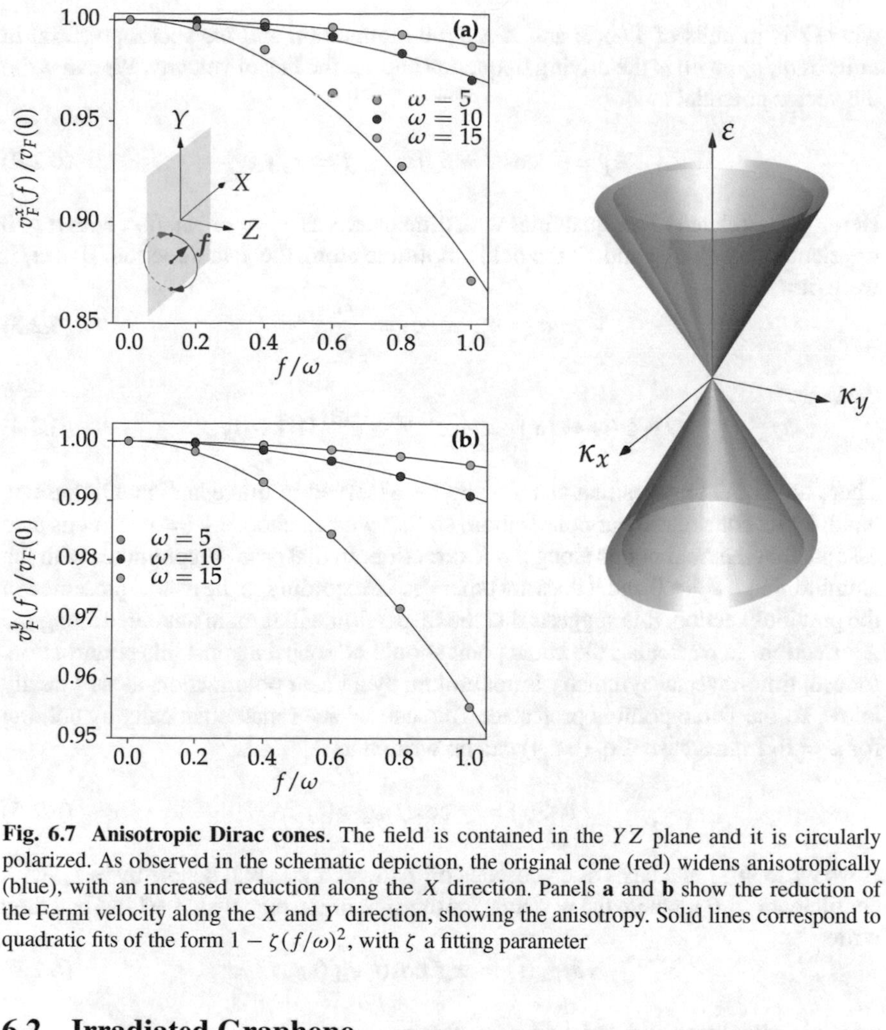

Fig. 6.7 Anisotropic Dirac cones. The field is contained in the YZ plane and it is circularly polarized. As observed in the schematic depiction, the original cone (red) widens anisotropically (blue), with an increased reduction along the X direction. Panels **a** and **b** show the reduction of the Fermi velocity along the X and Y direction, showing the anisotropy. Solid lines correspond to quadratic fits of the form $1 - \zeta(f/\omega)^2$, with ζ a fitting parameter

6.2 Irradiated Graphene

Although graphene has been studied extensively in the literature [8, 10, 13, 15, 16, 20, 22], we believe it to be useful to discuss briefly some of its properties under driving. Indeed, we expect to observe a similar behaviour as in the topological boundary. Namely, a reduction of the Fermi velocity both in linear and circular polarizations, together with a gap opening at the Dirac point. In the bulk, we may consider the two valleys to be uncoupled so we may focus on a single valley and write the Weyl equation for irradiated graphene as

$$\mathrm{i}\,\partial_t \boldsymbol{\Phi}(t) = \boldsymbol{\sigma}_\perp \cdot [\boldsymbol{\kappa} + \boldsymbol{A}(t)]\, \boldsymbol{\Phi}(t)\,, \tag{6.2.1}$$

where t is in units of $1/\omega$, $\boldsymbol{\kappa}$ and $\boldsymbol{A}$ are the momentum and the vector potential in units of ω/v_F, with ω the driving frequency and v_F the Fermi velocity. We can write the vector potential as

$$A_j = f_j \cos(t + \theta_j) , \qquad j = x, y . \tag{6.2.2}$$

Here, f_j are related to the quantities with dimensions as $f_j = eF_j v_F/\omega^2$, with $e > 0$ the elementary charge and F_j the field amplitude along the j-th direction. Hence, if we write

$$V = \boldsymbol{\sigma}_\perp \cdot \boldsymbol{a} , \qquad a_j = \frac{f_j}{2} e^{\mathrm{i}\theta_j} , \tag{6.2.3}$$

then

$$\mathrm{i}\,\partial_t \boldsymbol{\Phi}(t) = \left[\boldsymbol{\sigma}_\perp \cdot \boldsymbol{\kappa} + e^{\mathrm{i}t} V + e^{-\mathrm{i}t} V^\dagger\right] \boldsymbol{\Phi}(t) . \tag{6.2.4}$$

There are some features that can already be observed in this equation. On the one hand, let us consider linear polarization, so that we can choose $V = V^\dagger$. Let us also assume that the field points along the X direction. In that case, V commutes with the Hamiltonian if $\kappa_y = 0$ and it does not otherwise. According to the results presented in the previous section, this suggests that the dispersion will remain unaltered along the X direction. In particular, the Dirac point should be robust against this perturbation. Indeed, time-reversal symmetry is not broken by a linear polarization, as we already know, so the Dirac point is protected. This can be seen mathematically by solving for $\boldsymbol{\kappa} = 0$. In that case, Eq. (6.2.4) can be written as

$$\mathrm{i}\,\partial_t \boldsymbol{\Phi}(t) = f \cos(t) \sigma_x \boldsymbol{\Phi}(t) . \tag{6.2.5}$$

If we write $\boldsymbol{\Phi}(t)$ in a basis of eigenstates of σ_x, which we shall denote by $|\pm\rangle_x$, with coefficients $\varphi_\pm(t)$, where the $\pm$ stands for the sign of the eigenvalue ± 1, we can then write

$$\mathrm{i}\,\partial_t \varphi_\pm(t) = \pm f \cos(t) \varphi_\pm(t) , \tag{6.2.6}$$

which is trivially solved and we finally obtain

$$\boldsymbol{\Phi}(t) = c_+ \exp\left[-\mathrm{i} f \sin(t)\right] |+\rangle_x + c_- \exp\left[\mathrm{i} f \sin(t)\right] |-\rangle_x , \tag{6.2.7}$$

where $c_\pm$ are integration constants. We see that $\boldsymbol{\Phi}(t + 2\pi) = \boldsymbol{\Phi}(t)$. However, according to Floquet's theorem, $\boldsymbol{\Phi}(t + 2\pi) = \exp(-\mathrm{i}\,2\pi\,\varepsilon)\boldsymbol{\Phi}(t)$. Combining both conditions we obtain that $\varepsilon = l$, with l an integer. Here ε is the quasienergy in units of ω. As we can see, there is no gap opening, as predicted, regardless of the intensity of the field.

In the case of circularly polarized light, we can write Eq. (6.2.4) as follows

$$\mathrm{i}\,\partial_t \boldsymbol{\Phi}(t) = f\left[e^{\mathrm{i}t} \sigma_+ + e^{-\mathrm{i}t} \sigma_-\right] \boldsymbol{\Phi}(t) , \tag{6.2.8}$$

with $\sigma_{\pm} = (\sigma_x \pm \mathrm{i}\,\sigma_y)/2$. In order to solve this equation, we perform a rotation of angle t about the Z axis with

$$\mathcal{R}_z(t) = \exp\left[\mathrm{i}\,\sigma_z(t/2)\right] . \tag{6.2.9}$$

Hence, if we introduce

$$\boldsymbol{\Phi}(t) = \mathcal{R}_z(t)\boldsymbol{\Psi}(t) , \tag{6.2.10}$$

into Eq. (6.2.8), we obtain

$$\mathrm{i}\,\partial_t \boldsymbol{\Psi}(t) = (\sigma_z/2 + f\sigma_x)\,\boldsymbol{\Psi}(t) . \tag{6.2.11}$$

This equation can be solved using the matrix exponential

$$\boldsymbol{\Psi}(t) = \exp\left[-\mathrm{i}\,(\sigma_z/2 + f\sigma_x)\,t\right]\boldsymbol{\varphi} , \tag{6.2.12}$$

with $\boldsymbol{\varphi}$ a constant vector. We now need to impose the conditions of Floquet's theorem. However, careful must be taken since a rotation of 2π carries a minus sign, that is, $\mathcal{R}_z(\theta + 2\pi) = -\mathcal{R}_z(\theta)$. This in turn implies that

$$\boldsymbol{\Psi}(t + 2\pi) = -e^{-\mathrm{i}2\pi\varepsilon}\,\boldsymbol{\Psi}(t) . \tag{6.2.13}$$

If we take into account that the eigenvalues of the matrix inside the exponential are given by $\pm\sqrt{1+4f^2}/2$, we immediately find

$$\varepsilon_n^{\pm} = \pm\frac{1}{2}\sqrt{1+4f^2} + n_{\pm} + \frac{1}{2}, \qquad n_{\pm} \in \mathbb{Z} . \tag{6.2.14}$$

If $f^2 < 3/4$, the two branches within the first Floquet-Brillouin zone will be

$$\varepsilon_{\pm} = \pm\frac{1}{2}\left(\sqrt{1+4f^2} - 1\right) . \tag{6.2.15}$$

Hence, there is a quasienergy gap

$$2\delta = \sqrt{1+(2f)^2} - 1 . \tag{6.2.16}$$

Interestingly enough, for low enough fields $\Delta \propto f^2$, just like in the previous section. We will show later that this is indeed the case when numerically solving the problem. Notice that the gap increases with the field until the two branches hit the edges of the Floquet-Brillouin zone when $f^2 = 3/4$. Upon increasing the field further and in order to remain within the Floquet-Brillouin zone, the two branches move towards zero quasienergy, closing up the gap when $f^2 = 2$. Alternatively, one can think that the two original branches move into the nearest Floquet-Brillouin zones, while branches from those neighbouring zones get inside the first one, thereby reducing the

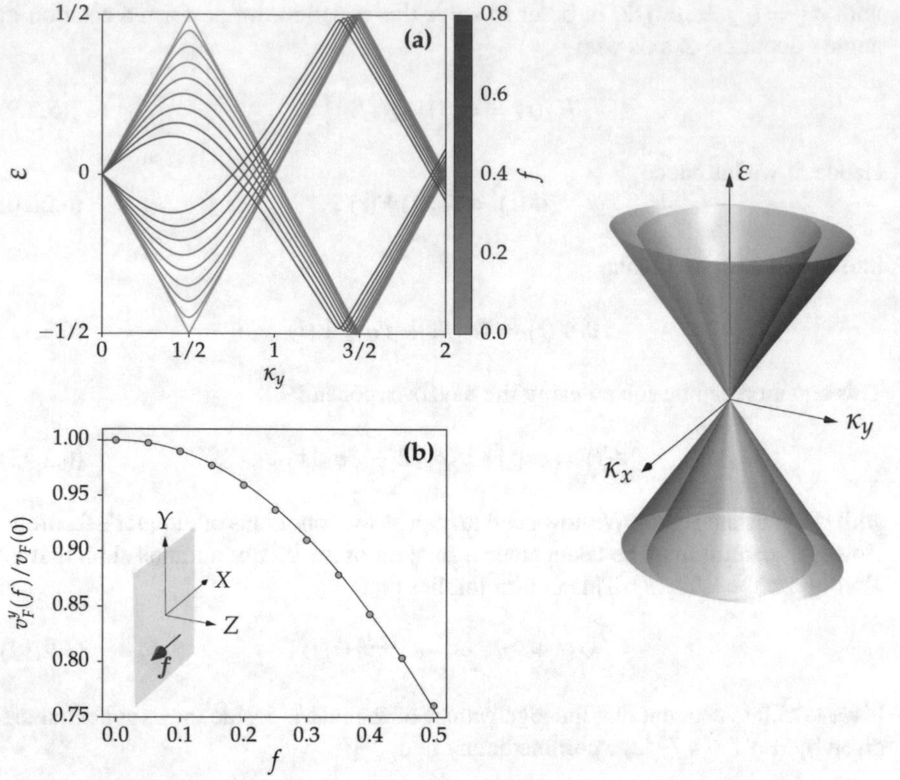

Fig. 6.8 Anisotropic Dirac cones. The field is in-plane and linearly polarized. As observed in the schematic depiction, the original cone (red) widens in the direction perpendicular to the field to form an anisotropic cone (blue). **(a)** Dispersion for different values of the field as indicated by the color bar. As long as the field is nonzero, there are avoided crossings at the edge of the Floquet-Brillouin zone and the slope for low momenta gets reduced. **(b)** Velocity reduction together with a quadratic fit (solid line) of the form $1-\zeta(f/\omega)^2$, with ζ a fitting parameter

gap until it becomes zero. Then, the process starts again. Hence, touching the edges of the Brillouin zone occurs whenever $\sqrt{1+4f^2}$ becomes an even integer, that is, when $f^2 = m^2 - 1/4$, with $m \in \mathbb{Z}^+$. On the other hand, the gap closes whenever $\sqrt{1+4f^2}$ becomes an odd integer, that is, when $f^2 = m(m-1)$, with $m \in \mathbb{Z}^+$.

In order to obtain the dispersion with $\boldsymbol{\kappa}$, we can proceed as in the previous section. In that case, using Floquet's theorem, Eq. (6.2.4) is straightforwardly written in terms of the Fourier components as

$$\varepsilon\,\boldsymbol{\varphi}_l = [\boldsymbol{\sigma}\cdot\boldsymbol{k} - l\mathbb{1}_2]\,\boldsymbol{\varphi}_l + V\boldsymbol{\varphi}_{l+1} + V^\dagger\boldsymbol{\varphi}_{l-1}\,. \tag{6.2.17}$$

This problem is similar to a nearest-neighbours tight-binding model with two degrees of freedom per site and a site-dependent energy. In contrast to the topological

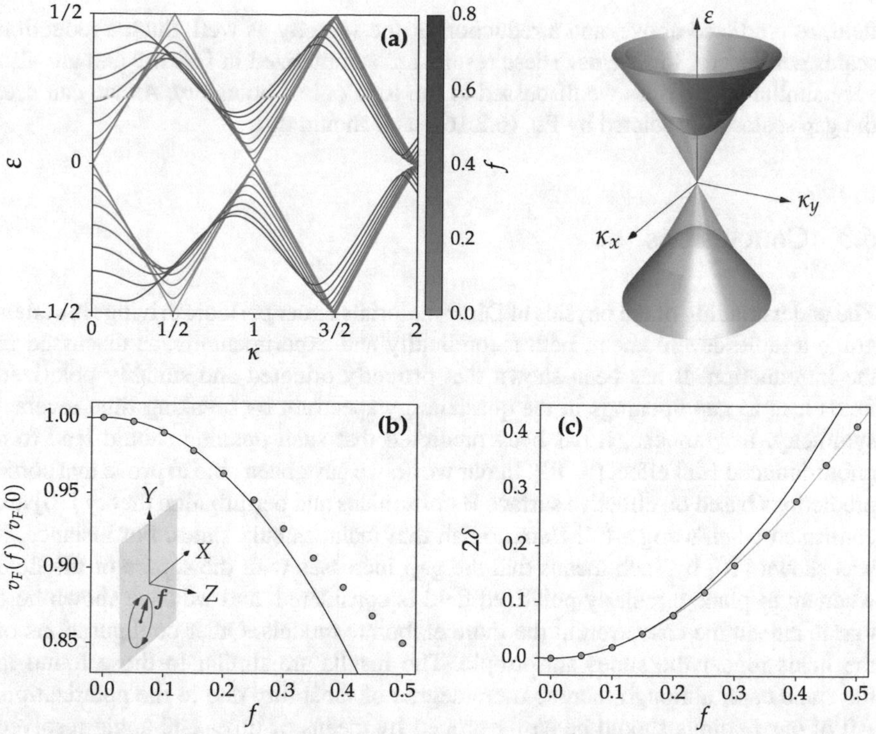

Fig. 6.9 **Massive Dirac dispersion**. The field is in-plane and circularly polarized. As observed in the schematic depiction, the original cone (red) widens isotropically and a gap opens up at the Dirac point to form a massive Dirac dispersion (blue). **a** Dispersion for different values of the field as indicated by the color bar. As long as the field is nonzero, a gap opens up and increases in size with the field, while reducing the slope of the dispersion for low momenta. **b** Velocity reduction together with a quadratic fit (solid line) for low fields of the form $1 - \zeta(f/\omega)^2$, with ζ a fitting parameter. **c** Quasienergy gap, together with the analytic result of Eq. (6.2.16) [orange line] and the quadratic approximation for low fields

boundary, this equation is easily solved numerically since it corresponds directly to a tridiagonal block matrix and we do not have to discretize along any direction. If we consider a linearly polarized field with $f_x = f$ and $f_y = 0$, we find that the dispersion is unaffected along the X-direction. In contrast, it gets widened along the Y-direction and the Dirac point is preserved, as we expected. Moreover, the velocity decreases in good agreement with a reduction that scales with f^2, similarly to what we obtained in the previous section for the topological boundary. These results are summarized in Fig. 6.8. The sideband index has been limited to 10 on the basis that we focus on low fields. As we can see, there is indeed a reduction on the velocity for low momenta and avoided crossings occur at the edges of the Brillouin zone. The results are very similar to those in the topological boundary. If we consider circular polarization instead, there is a gap opening at the Dirac point that squares with the field, as predicted above, and a reduction of the velocity as well. Such a reduction scales with f^2 for low fields. These results are summarized in Fig. 6.9 and are also very similar to the ones we discussed in the topological boundary. As we can see, the gap scales as predicted by Eq. (6.2.16), as it should do.

6.3 Conclusions

The understanding of the physics of Dirac materials under periodic drivings is undergoing a rapid development, both theoretically and experimentally, as discussed in the introduction. It has been shown that properly oriented and suitably polarized fields lead to gap openings in the quasienergy spectrum by breaking time-reversal symmetry. In graphene, it has been predicted that such opening should lead to a photo-induced Hall effect [8, 10]. In our work, we have been able to prove that some predictions based on effective surface Hamiltonians and perturbation theory [10] are confirmed when using a full Hamiltonian that includes bulk states. For instance, it was shown [10] by such means that the gap increases with the square of the field when an in-plane circularly polarized field is considered, and we have shown here that is indeed the case even in the more elaborate models. Other configurations of the fields render the cones anisotropic. The results are similar to those found in the static case, although there is more degree of tunability due to the polarization. All of our findings should be easily probed by means of time- and angle-resolved photoemission spectroscopy, as discussed in Ref. [17].

References

1. Eckardt A (2017) Colloquium: atomic quantum gases in periodically driven optical lattices. Rev Mod Phys 89:011004
2. Fleury R, Khanikaev AB, Al A.: u. Floquet topological insulators for sound. Nat Commun 7: 11744 (2016)
3. Rechtsman MC, Zeuner JM, Plotnik Y, Lumer Y, Podolsky D, Dreisow F, Nolte S, Segev M, Szameit A (2013) Photonic Floquet topological insulators. Nature 496:196
4. Floquet G (1883) Sur les équations différentielles linéaires ‘a coefficients périodiques. Ann Sci Ecole Norm S 12:47
5. Zel’dovich YB (1967) The quasienergy of a quantum-mechanical system subjected to a periodic action. Sov Phys JETP 24:1006
6. Grifoni M, Hänggi P (1998) Driven quantum tunneling. Phys Rep 304:229
7. Platero G, Aguado R (2004) Photon-assisted transport in semiconductor nanostructures. Phys Rep 395:1
8. Oka T, Aoki H (2009) Photovoltaic Hall effect in graphene. Phys Rev B 79:081406
9. Kitagawa T, Berg E, Rudner M, Demler E (2010) Topological characterization of periodically driven quantum systems. Phys Rev B 82:235114
10. Kitagawa T, Oka T, Brataas A, Fu L, Demler E (2011) Transport properties of nonequilibrium systems under the application of light: photoinduced quantum Hall insulators without Landau levels. Phys Rev B 84:235108

11. Lindner NH, Refael G, Galitski V (2011) Floquet topological insulator in semiconductor quantum wells. Nat Phys 7:490
12. Gómez-León A, Platero G (2013) Floquet-bloch theory and topology in periodically driven lattices. Phys Rev Lett 110:200403
13. Delplace P, Gómez-León A, Platero G (2013) Merging of Dirac points and Floquet topological transitions in ac-driven graphene. Phys Rev B 88:245422
14. Nathan F, Rudner MS (2015) Topological singularities and the general classification of Floquet-Bloch systems. New J Phys 17:125014
15. Syzranov SV, Rodionov Ya I, Kugel KI, Nori F (2013) Strongly anisotropic Dirac quasiparticles in irradiated graphene. Phys Rev B 88:241112
16. Usaj G, Perez-Piskunow PM, Torres LEF, Balseiro CA (2014) Irradiated graphene as a tunable Floquet topological insulator . Phys Rev B 90: 115423
17. Wang YH, Steinberg H, Jarillo-Herrero P, Gedik N (2013) Observation of Floquet-Bloch States on the surface of a topological insulator. Science 342:453
18. Díaz-Fernández A, Díaz E, Gómez-León A, Platero G, Domínguez- Adame F (2019) Floquet engineering of Dirac cones on the surface of a topological insulator. Phys Rev B 100:075412
19. Díaz-Fernández A (2020) Inducing anisotropies in Dirac fermions by periodic driving. J Phys: Condens Matter 32:495501
20. Farrell A, Arsenault A, Pereg-Barnea T (2016) Dirac cones, Floquet side bands, and theory of time-resolved angle-resolved photoemission. Phys Rev B 94:155304
21. Díaz E, Miralles K, Domínguez-Adame F, Gaul C (2014) Spin dependent terahertz oscillator based on hybrid graphene superlattices. Appl Phys Lett 105:103109
22. Agarwala A, Bhattacharya U, Dutta A, Sen D (2016) Effects of periodic kicking on dispersion and wave packet dynamics in graphene. Phys Rev B 93:174301

Chapter 7
Conclusions

In this Thesis, three-dimensional topological insulators and graphene are exposed to perturbations, in order to explore the consequences of such perturbations on the massless Dirac excitations of both systems. This chapter is devoted to gather concluding remarks pointed out on each chapter, in order to restate the main results of this Thesis.

The most salient feature of all chapters is the robustness of the Dirac point when external perturbations that preserve the symmetries are applied. That is, topological protection does indeed work when the system is exposed to the action of electric and magnetic fields. In the case of electric fields [cf. Chap. 3], we have been able to show that a topological boundary and graphene metallic armchair nanoribbons lower their Fermi velocity as the field strength is increased. Importantly, such a reduction cannot be captured by means of first order perturbation theory. Our results seem generic to other Dirac materials and should have important consequences on quantum transport, since the Fermi velocity is a crucial parameter to this regard. It is also interesting to point out that this reduction can also be exploited to lower the energy gap that arises at a thin film due to annihilation of opposite helicities. When a magnetic field is applied perpendicular to a topological boundary, the Dirac cones evolve into a set of relativistic Landau levels. Since the Fermi velocity enters directly into the spacing between Landau levels in relativistic quantum materials, upon applying both electric and magnetic fields one should be able to control such a spacing. Therefore, a change in the Fermi velocity would be measurable by magneto-transport techniques, apart from the obvious angle-resolved photoemission spectroscopy.

If the magnetic field respects mirror symmetry, even though it breaks time-reversal symmetry, the topological insulator may host a topological crystalline insulating phase [cf. Chap. 4]. This is exactly what occurs in the materials of our interest when the magnetic field is contained within the topological boundary. We can observe that, due to the presence of the topological boundary, the Landau orbits close to the boundary become dispersive in the direction perpendicular to the magnetic field. This is in contrast to usual bulk systems, where Landau levels are only dispersive along the direction parallel to the field. Moreover, the velocity in the direction per-

Á. Díaz Fernández, *Reshaping of Dirac Cones in Topological Insulators and Graphene*, Springer Theses, https://doi.org/10.1007/978-3-030-61555-0_7

pendicular to the field is reduced with respect to the bare Fermi velocity, rendering an anisotropic cone for low momenta. When an additional electric field perpendicular to the boundary is applied, a local potential is experienced by the Landau levels and their dispersion is notably anisotropic, while still preserving a massless Dirac-like behaviour.

On a different note, topological surface states are said to be robust against disorder. One way to observe if this is indeed true is by cleaving a topological insulator and exposing it to different environments. Experiments have shown that the topological surface state persists. Moreover, this surface state coexists with a two-dimensional Rashba-split electron gas. The appearance of this electron gas takes place due to band-bending effects that result from the electric potential created by the ionized donor impurities and the electrons these have donated to the host material. As a result from such band-bending effects, a quantum well is formed for states in the continuum, which get sucked in by such a potential, leading to a quantization of the transverse momentum, thereby forming subbands. The reason for having Rashba splitting can be ascribed to the fact that there is structural inversion asymmetry, together with an effective in-built electric field. Although these contributions are truly relevant in proving a point, they are far from controllable. In this Thesis [cf. Chap. 5], we propose instead to evaporate a δ-layer of donor atoms during growth. A control of the density of impurities and thereby of the built-in electric field can be achieved with noteworthy precision. A solvable Thomas-Fermi approximation to obtain the built-in electric potential, along with a $3+1$ Dirac equation to model the topological insulators, allows us to obtain the Rashba-split spectrum analytically and to confirm the spin textures predicted for such systems. Moreover, we show that the oscillator strength, a measure of the optical absorption, gets largely reshaped by the presence of the topological surface state in contrast to the trivial scenario.

The last chapter of this Thesis [cf. Chap. 6] deals with the application of periodic drivings to a topological boundary and graphene. By means of Floquet theory and working in the high-frequency limit, we observe that different polarizations and orientations of the field lead to a reshaping of the Dirac cones in the quasienergy spectra. In particular, we obtain that an in-plane, circularly polarized field leads to a gap opening at the Dirac point due to time-reversal symmetry breaking. This result was known to occur by means of perturbation theory using effective surface Hamiltonians. We have been able to show that this phenomenon occurs even when the bulk levels are considered in the numerical solution of the problem. Additionally, our results show very similar behaviours as those observed in the static case. Namely, we observe a reduction of the slope of the cones that follows exactly the same trend as does the reduction of the Fermi velocity in the static case.

In summary, by means of external perturbations, we have been able to modify intrinsic properties of topological insulators and graphene, such as the Fermi velocity. Moreover, some topological properties ascribed to these materials have been confirmed, such as the robustness of the Dirac point in situations where time-reversal and mirror symmetry are preserved. This should in turn have an impact on quantum transport and we believe that experiments to measure the results presented herein should be accesible.

Appendix
Floquet Matrix in a Topological Boundary

In Chap. 6, we faced the following problem

$$\varepsilon\boldsymbol{\varphi}_l(z) = \mathcal{H}_l^0\boldsymbol{\varphi}_l(\xi) + J\boldsymbol{\varphi}_{l+1}(z) + J^\dagger\boldsymbol{\varphi}_{l-1}(z) \,, \tag{A.1}$$

with $J = \boldsymbol{\alpha}\cdot\boldsymbol{a}$ and

$$\mathcal{H}_l^0 = \alpha_z p_z + \boldsymbol{\alpha}_\perp\cdot\boldsymbol{\kappa} + \beta\,\mathrm{sgn}\,(z) - l\omega \,. \tag{A.2}$$

Details of the quantities above can be found in Chap. 6. Here l is the sideband index and runs from $-N_\omega$ to N_ω. Let us write Eq. (A.1) in component form. For that matter, one has to take into account that the only non-zero components of the Dirac matrices are

$$\begin{aligned}\alpha_x^{\mu,3-\mu} &= 1\,, & \alpha_y^{\mu,3-\mu} &= (-1)^{\mu+1}\,\mathrm{i}\\ \alpha_z^{\mu,(\mu+2)\bmod 4} &= (-1)^\mu\,, & \beta^{\mu,\mu} &= (-1)^{\mathrm{int}(\mu/2)}\,,\end{aligned} \tag{A.3}$$

with $\mu = 0, 1, 2$ and 3. Hence, Eq. (A.1) can be written as

$$\begin{aligned}&\left[\varepsilon - (-1)^{\mathrm{int}(\mu/2)}g(z-L_z) + l\omega - N_\omega\omega\right]\varphi_l^\mu(z)\\ &= p_x\varphi_l^{3-\mu}(z) + (-1)^{\mu+1}ip_y\varphi_l^{3-\mu}(z) + (-1)^\mu p_z\varphi_l^{(\mu+2)\bmod 4}(z)\\ &\quad + a_x\varphi_{l+1}^{3-\mu}(z) + (-1)^{\mu+1}ia_y\varphi_{l+1}^{3-\mu}(z) + (-1)^\mu a_z\varphi_{l+1}^{(\mu+2)\bmod 4}(z)\\ &\quad + a_x^*\varphi_{l-1}^{3-\mu}(z) + (-1)^{\mu+1}ia_y^*\varphi_{l-1}^{3-\mu}(z) + (-1)^\mu a_z^*\varphi_{l-1}^{(\mu+2)\bmod 4}(z)\,.\end{aligned} \tag{A.4}$$

Here, we have displaced the origin of the sideband index so that l runs from 0 to $2N_\omega$ and the origin in the position variable so that the system is inside a box from $z = 0$ to $z = 2L_z$. The z-dependence implies that $p_z = -\mathrm{i}\,\partial_z$. We will discretize on a lattice of spacing $d = L_z/N_z$, with $2N_z + 1$ the total number of sites. Instead of sampling all four components of the bispinor in all the lattice sites, we will alternate

Á. Díaz Fernández, *Reshaping of Dirac Cones in Topological Insulators and Graphene*, Springer Theses, https://doi.org/10.1007/978-3-030-61555-0

and sample φ^0 and φ^3 in the even lattice sites and φ^1 and φ^2 in the odd lattice sites, following reference [1]. In order to do so, we can do the following mapping

$$\varphi_l^\mu(jd) \to F_l^\alpha(jd)\,, \qquad \alpha = \mu \bmod 2\,, \tag{A.5}$$

where $j = 0, 1, \ldots, 2N_z$. Since α and μ have the same parity, all those terms with $(-1)^\mu$ are directly replaced by $(-1)^\alpha$. The mapping given in Eq. (A.5) implies that

$$\begin{aligned} \varphi_l^\mu(z) &\to F_{l,j}^\alpha\,, \qquad \varphi_l^{3-\mu}(z) = F_{l,j}^{1-\alpha}\,, \\ p_z\varphi_l^{(\mu+2)\bmod 4}(z) &\to -i\frac{1}{2d_z}\left[F_{l,j+1}^\alpha - F_{l,j-1}^\alpha\right]\,, \\ \varphi_l^{(\mu+2)\bmod 4}(z) &\to \frac{1}{2}\left[F_{l,j+1}^\alpha + F_{l,j-1}^\alpha\right]\,, \end{aligned} \tag{A.6}$$

where $F_{l,j}^\alpha = F_l^\alpha(jd_z)$. Here we have taken into account that the $(\mu+2) \bmod 4$ components are alternated with the μ components, so that if μ is sampled in j, then $(\mu+2) \bmod 4$ is sampled in $j+1$ and $j-1$. Hence, the last term is an interpolation of these two. However, since $(\mu+2) \bmod 4$ shares the same parity as μ, both have the same α. In contrast, $3-\mu$ has opposite parity to μ, so that if μ maps to α, $3-\mu$ maps to $1-\alpha$. The only term left to deal with is

$$(-1)^{\mathrm{int}(\mu/2)}\,. \tag{A.7}$$

For this term, we can see that $\mu = 0$ and 1 lead to a $+1$ whereas $\mu = 2$ and 3 lead to a -1. On the other hand, $\mu = 0$ and $\mu = 3$ are sampled in the even lattice sites, whereas $\mu = 1$ and $\mu = 2$ are sampled in the odd lattice sites. This means that $j+\alpha$ is even for $\mu = 0$ and 1 and for $\mu = 2$ and 3 we have that $j+\alpha$ is odd. Therefore, we can do the following

$$(-1)^{\mathrm{int}(\mu/2)} \to (-1)^{j+\alpha}\,. \tag{A.8}$$

Since

$$z - L_z = jd_z - L_z = \left[\frac{j}{N_z} - 1\right] L_z = (j - N_z)\frac{L_z}{N_z}\,, \tag{A.9}$$

we can write

$$g(z - L_z) = g(j - N_z) \equiv g_j\,, \tag{A.10}$$

since g is the sign function and L_z/N_z is always a positive number. Finally, we can write Eq. (A.4) as follows

$$\begin{aligned}
&\Big[\varepsilon-(-1)^{j+\alpha}g_j+l\omega-N_\omega\omega\Big]F_{l,j}^{\alpha}\\
&\quad=p_xF_{l,j}^{1-\alpha}+(-1)^{\alpha+1}ip_yF_{l,j}^{1-\alpha}+(-1)^{\alpha+1}\frac{i}{2d}\left[F_{l,j+1}^{\alpha}-F_{l,j-1}^{\alpha}\right]\\
&\qquad+a_xF_{l+1,j}^{1-\alpha}+(-1)^{\alpha+1}ia_yF_{l+1,j}^{1-\alpha}+(-1)^{\alpha}\frac{a_z}{2}\left[F_{l+1,j+1}^{\alpha}+F_{l+1,j-1}^{\alpha}\right]\\
&\qquad+a_x^*F_{l-1,j}^{1-\alpha}+(-1)^{\alpha+1}ia_y^*F_{l-1,j}^{1-\alpha}+(-1)^{\alpha}\frac{a_z^*}{2}\left[F_{l-1,j+1}^{\alpha}+F_{l-1,j-1}^{\alpha}\right].
\end{aligned} \tag{A.11}$$

We have three indices, $\alpha = 0, 1$, $j = 0, \dots, 2N_z$ and $l = 0, \dots, 2N_\omega$. By joining all of those into a single index one can write a matrix for its later diagonalization to obtain the quasienergies, ε.

Reference

1. Díaz E, Miralles K, Domínguez-Adame F, Gaul C (2014) Spin dependent terahertz oscillator based on hybrid graphene superlattices. Appl Phys Lett 105:103109

Printed in the United States
by Baker & Taylor Publisher Services

Printed in the United States
by Baker & Taylor Publisher Services